POLITICAL ECOLOGY

POLITICAL ECOLOGY

*An Integrative Approach to Geography
and Environment–Development Studies*

Edited by
KARL S. ZIMMERER
and
THOMAS J. BASSETT

THE GUILFORD PRESS
New York   London

© 2003 The Guilford Press
A Division of Guilford Publications, Inc.
72 Spring Street, New York, NY 10012
www.guilford.com

Printed in the United States of America

This book is printed on acid-free paper.

Last digit is print number: 9   8   7   6   5   4   3   2   1

Library of Congress Cataloging-in-Publication Data

Political ecology : an integrative approach to geography and environment–
development studies / edited by Karl S. Zimmerer and Thomas
   J. Bassett.
      p.   cm.
   Includes bibliographical references and index (p.   ).
   ISBN 1-57230-916-4 (pbk.: alk. paper)
   1. Political ecology.   I. Zimmerer, Karl S.   II. Bassett, Thomas J.
   JA75.8.P63 2003
   04.2—dc22

                                                                    2003016121

Acknowledgments

This project is a collaborative effort that began nearly 5 years ago. As former chairs of the Cultural Ecology Specialty Group of the Association of American Geographers (a large theme group within the main U.S. geographical organization), we were commissioned to write a state-of-the-art review and analysis of cultural and political ecology for the association's *Benchmark, 2000* volume. We have considered hundreds of books and articles during the past several years, and are grateful to the various people who provided us with materials, advice, and feedback. In particular we thank Morgan Robertson (Madison) and Rob Daniels (Urbana–Champaign), who aided in our extensive library research.

Our review convinced us of the need for an in-depth and thematically focused presentation of the dynamic geographical approach to political ecology, which became the focus of this book. This rapidly growing approach to environment and development issues has guided our own research and teaching for a combined total of more than 30 years. We share a special interest in approaching political ecology as geographers. This interest has led us to include chapters that identify and advance some core topics and themes of this approach to political ecology. For their assistance in doing so, we are especially grateful to the authors for their enthusiasm and cooperation and to Kristal Hawkins, our editor at The Guilford Press. We also wish to thank the Cartography Laboratory of the University of Wisconsin–Madison, especially its Director, Onno Brouwer, and cartographers Caitlin Doran and Rich Worthington, for the fine graphics production.

Contents

POLITICAL ECOLOGY

CHAPTER 1

Approaching Political Ecology

Society, Nature, and Scale
in Human–Environment Studies

Karl S. Zimmerer
Thomas J. Bassett

The field of political ecology has produced a number of high-quality but dispersed studies that have significantly contributed to our understanding of nature and society relations from a geographical perspective. This volume presents a group of key works that represent recent advances, debates, and the ongoing evolution of political ecology. Our collection seeks to advance political ecological studies in four areas. First, it strives to engage both the ecological and the political dimensions of environmental issues in a more balanced and integrated manner. Up until now, most political ecological research emphasized just one of these dimensions. Rarely does one find a synthesis of the political and ecological processes whose outcomes produce distinctive problems and suggest particular solutions. Second, this book expands the geographical range of political ecological studies to include urban and industrial settings of the global North and South. Heretofore, most political ecological research has had a rural and developing-world focus. In this book we promote the extension of political ecology to these new settings and encourage the historical study of North–South environmental relationships. Third, this book strengthens one of the analytical cores of political ecology by arguing for a more creative consideration of geographical scale. The scales of social and ecological processes and their interactive effects on environmental problems and policies have received growing attention among political ecologists. We view scalar politics and ecological scales as important dimensions of environmental change, and we urge political ecologists to explore their dynamic interplay further. A fourth and final contribution of this collection is its sustained discussion of research methods in political ecology. How do we collect the materials that allow us to construct our arguments? What is the correct mix of quantitative and qualitative methods (and how do we do it)? The authors of each chapter are explicit about their methodological approaches to their topics. Collectively, they serve as a primer of sorts for scholars and those interested in undertaking research in a geography-centered political ecology.

The general goal of this volume, then, is consolidating and contributing to advances in what we consider to be a particularly productive branch of political ecology that is centered around the themes of nature–society interaction and geographical scale. Our volume is definitely *not* a reader that would attempt to encompass the broad terrain of political ecology in its characteristically interdisciplinary, highly diverse, and perhaps even increasingly divergent forms. To show this, we have titled the book *Political Ecology: An Integrative Approach to Geography and Environment–Development Studies* in order to distinguish our geographical approach from the complementary approaches rooted in anthropology, sociology, history, forestry, political science, and studies of technology and science (Painter & Durham, 1995; Fairhead & Leach, 1996; Peluso, 1992; Schmink & Wood, 1987). We recognize also that our approach in this book can be thought of as *a* geographical political ecology, that is, as *one* of several current approaches.

By framing our volume as a political ecology, we are highlighting our commitment to the diversity and plurality of political ecology. We consciously seek to avoid disciplining political ecology in the sense of excluding complementary and equally vibrant approaches. To the contrary, we encourage the flourishing of political ecology to include the fullest possible range of approaches. Indeed, our geographical approach strongly endorses the tight interweaving of disciplines that is a defining characteristic of political ecology. The last section of this chapter introduces our contributions to the issues that are raised of interdisciplinarity (the linking and moderate integration of disciplines) and transdisciplinarity (the high-level, fused integration of disciplines).

At the same time, our volume is committed to building upon the productive tension that exists between political ecology and the disciplines (that of geography in our case). With the rapid growth of political ecology and its increased engagement with several of the mainstream disciplines, we argue that the time is ripe for further exploring this productive tension between disciplinary contributions and the broad inter- and transdisciplinary field of inquiry that is the heart of political ecology. Admittedly, our volume develops a particular focus. Still, our desire is to see this focus as part of a dialogue that we hope will be of broad relevance to those interested in the other variants of political ecology, including those that may be more rooted in the other branches of geography, sociology, anthropology, planning, environmental history, and environmental resource management studies.

CORE THEMES

The geographical approach to political ecology featured in this book centers around two themes: social–environmental interactions and the political ecology of scale. The first theme contrasts with the "environmental politics" or "politicized environment" approaches that dominate current political ecology texts (Bryant & Bailey, 1997; Peet & Watts, 1996; Stott & Sullivan, 2000). *Political Ecology: An Integrative Approach to Geography and Environment–Development Studies* seriously engages with the biophysical as well as the social worlds. That is, the authors view the environment not simply as a stage or arena in which struggles over resource access and control take place. We consider nature, or biophysical processes, to play an active role in shaping human–environmental dynamics. Evaluating these processes is dependent on the use of the concepts and analysis of environmental science, such as ecology and physical geography. The latter approaches are used to examine biogeophysical factors and processes. At the same time, how these processes are chosen for study and how we eventually explain human–environmental interactions takes us to the politics and culture of representations of "nature" and the narratives that give them form and meaning (Bryant, 1998; Cronon, 1992, 1996; Slater, 2000). This philosophical embrace of the environment as having an ontological basis and a dynamic role as an agent in its own right, combined with our understanding of nature's agency as socially mediated, reflects a "natural turn" in the social sciences that is known as "critical realism" (Eden, 2001, p. 83). One of this book's goals is to elaborate what we consider to be a series of key works distinguishing the focus on the "natural turn" in political ecology.

A second goal is to demonstrate the centrality of geographical scale to political ecological analysis. This second organizing theme, the political ecology of scale, is closely entwined with the nature–society core focus of this book. Simply put, diverse environmental processes interact with social processes, creating different scales of mutual relations that produce distinctive political ecologies. The focus on scale builds on Piers Blaikie's original emphasis on multiscale political economic processes affecting local resource use patterns. Yet the scalar configuration of what Blaikie called the "chain of explanation" conceptualized scale in a hierarchical fashion. Implicit in the chain model is a conceptualization of scale as a series of pregiven sociospatial containers such as rural–urban, local, regional, national, and international. Building on recent theoretical advances, this book considers scale not as ontologically given but as social-environmentally produced. The various case studies explore how political–ecological processes incorporate and generate scaled spaces of interaction. These studies incorporate new ideas from the social construction of scale literature and the burgeoning field of landscape ecology to highlight the relational and simultaneous nature of human–environmental scales (Brandt, 1999; Delaney & Leitner, 1997; Fry, 2001; Marston, 2000; Naveh, 1991, 2000; Swyngedouw, 1997a; Wiens, 1989). They demonstrate a variety of scalar configurations that display vertical (hierarchical, nested) and horizontal (networked) patterns (Jonas, 1994). This book seeks to contribute to the politics of scale literature by demonstrating the central importance of ecological scale in shaping political–ecological dynamics.

In pursuing its two core themes, this book also puts a primary emphasis on the role of geographic differences in political ecology. *Geographic difference* refers to the biogeophysical and social characteristics that are associated with the environments of varied places. The diverse places that are taken under consideration in political ecology studies traverse the spectrum from rural farms and villages, to city neighborhoods and factories, to boardrooms and office suites. Geographic differences exist among local places and they are also, of course, characteristic of contrasts at the regional, national, and international levels. Not surprisingly, geography and other fields that specialize in environmental studies are nowadays quite interested in the role of geographic differences. As geographers and political ecologists, we are especially aware of the powerful influence of places in shaping environmental issues, whether urban, rural, or those of the global North or South. Equally important, and pertinent to this volume, is that the significance of places to political ecology is subject to change due to the dynamic nature of the environmental world and our intellectual frameworks for understanding it.

The significance of geographic differences in political ecology is especially relevant to our volume due to recent and ongoing processes of environmental globalization. *Globalization* refers to the functional integration of internationally dispersed activities. Environmental globalization is spurred by the integration of planetary support services such as climate and vegetation, as well as by the modification of these global processes though global economic, political, and cultural changes. Since recent and ongoing globalization is a fundamental change in the nature of the environmental world and our ways of understanding that world, it is bound to alter the meaning of geographic difference. One view of globalization and environment that holds special relevance for our volume is what is called *ecological modernization* (Buttel, 2000), a perspective that recognizes the enlarged importance of environmental issues and, reasoning from this recognition, sees environmental management as central to the overall workings of present-day and future societies.

The ecological modernization view is based on awareness of the legal, institutional, and eco-industrial and technomanagerial reforms, such as laws for water and air pollution control, that have led, in varying degrees, to improvements in environmental quality. This shift is agreed upon as a partial reorientation of environmental thought in the social sciences, yet much uncertainty surrounds its geographic underpinnings and political ecological significance (Blaikie, 1998). For example, its overall message remains murky or at least prematurely optimistic with regard to the role of geographic differences in the modernizing environmental management that is part of globalization; indeed, the claims for ecological modernization seem centered mostly on Northern Europe. Yet modernization-like changes concerned with such issues as national environmental action plans, biodiversity conservation, and protected areas are, at first glance, becoming as widespread in the global South as in the global North.

This collection features some dozen scholarly works that contribute to core themes underpinning our geographical political ecological approach. Most of these works were previously published but have been extensively revised for this book. The chapter by McCusker and Weiner was written especially for our volume. We have selected these works for their contribution to geographical political ecology in five substantive areas: Protected Areas and Conservation, Urban and Industrial Environments, Ecological Analysis and Theory in Resource Management and Conservation, Geospatial Technologies and Knowledges, and North–South Environmental Histories.

PART I. PROTECTED AREAS AND CONSERVATION

Protected areas occupy a large domain of political ecological research. The prominence of spatially defined conservation units (national parks, world heritage sites, wildlife corridors, biosphere reserves) has drawn geographers to examine the effects of these scaled spaces on access to and control of resources (Daniels & Bassett, 2002; Neumann, 1998; Schroeder, 1999; Zimmerer, 2000). The attempt to regulate resource use through controlling access is typically undertaken by delimiting conservation spaces and limiting use of heretofore common property resources. Political ecologists reveal how these spaces of conservation become arenas of conflict that result in distinctive patterns of resource management. Stevens (1997) writes on how the widely influential Yellowstone Model, which prohibits local groups from using protected resources, has giving way to alternative conservation strategies involving a variety of "participatory local management" approaches. His fieldwork in protected areas of Nepal shows that local management can be a viable basis for conservation and development (Stevens, 1993a, 1993b). Longitudinal research among indigenous peoples of Central America demonstrates both the feasibility and the challenges of grounding environmental conservation on the empowerment and participation of communities (Herlihy, 1992, 1993, 1997; Nietschman, 1973, 1997). The chapters contributed by Young (Chapter 2) and

Sundberg (Chapter 3) in this collection illustrate how local participation in natural resource conservation is strongly mediated by a community's interactions with nonlocal actors such as national governments, transnational corporations, and international nongovernmental organizations (NGOs). Together they reveal that these "conservation encounters" can shape landscapes and livelihoods in contradictory ways.

In her study of recreational whale watching along the Pacific Coast of Baja Mexico, Emily H. Young assesses the viability of this form of ecotourism for marine conservation and development. Her study probes into the interacting local and national politics of community-based conservation of common-pool resources and their consequences for marine habitats and wildlife. Young shows how access to the commons has shifted under changing political–economic conditions of centralized management of coastal resources, the commercialization of *ejido* land, and the dominance of large foreign companies in the tourism sector. Under these circumstances, few economic benefits trickle down to coastal communities. Poverty, in turn, only heightens pressure on depleted inshore fisheries. The recent involvement of Mexican and United States-based environmental NGOs in conserving marine habitats and promoting greater participation of local communities in ecotourism may aid local communities in future negotiations over local access rights to marine resources.

Juanita Sundberg's chapter examines how land users in the Petén peninsula of Guatemala have responded to the conservation agendas of NGOs charged with implementing the state-supported Maya Biosphere Reserve. Her study shows that those groups who conformed to the essentialist views and environmental discourses of NGOs as living in harmony with nature were in a better position to gain access to project benefits. People considered to lack the NGO-constructed cultural ecological traits were denied access to NGO-sponsored programs. As a result, different groups changed their land use practices in order to appear more "authentic" and thus enhance their access to project resources. Sundberg argues that the human–environmental discourses of conservation groups shape local landscapes and identities in ways that favor certain social groups and practices over others, with important consequences for livelihoods.

Common themes running through both chapters are the recognition that "communities" are characterized as much by their heterogeneity as by their (uneasy) alliance around certain issues, and that NGOs are important yet ambivalent intermediaries linking local and global concerns over conservation and development. Methodologically, both Sundberg and Young rely upon participant observation, extensive interviews, and other primary sources (e.g., newspapers, project documents) to support their arguments. Both chapters contribute to geographical political ecology by their sustained focus on human–environmental interactions. The political ecology of scale is evident in the mobility of people (Guatemala) and whales (Mexico), the scalar politics of resource management schemes (biosphere reserves, community-based marine protected areas), and the effects of these mismatched scales on conservation activities. Together these chapters demonstrate the insights that are emerging from new research on the political ecology of conservation.

PART II. URBAN AND INDUSTRIAL ENVIRONMENTS

Urban and industrial environments, as well as suburban settings, are widely held, including by political ecology, as central to the core concerns of contemporary environmentalism, environmental change, and conservation (Buttel, 1987; Cronon, 1991; Gottlieb, 1993; Keil, Bell, Penz, & Fawcett, 1998; Lipietz, 1995; Robbins, Polderman, & Birkenholtz, 2001; Rome, 2001). For political ecology, the challenge is pointed. How can this approach be useful to understanding the environments of variously urbanized and industrialized landscapes, which not only characterize the countries of Europe, Canada, and the United States but also predominant in most newly industrialized and developing countries. These settings are very different from rural locales, yet certainly they are neither fully disconnected nor entirely distinct (McCarthy, 2002; Myers, 1999; Robbins, 2002; Walker, 2003).

Mark Pelling's chapter (Chapter 4) further develops the political ecological analysis of urban and industrial environments. His case study examines the political ecology of risk that is associated with flooding in the capital city of Georgetown, located on the flood-prone coastal lowlands of Guyana. The study is timely, given the acute hazards of flooding in northern South America and the greater Caribbean (Pelling, 1997), and the likelihood that such risks will likely increase with greater storminess and higher sea levels brought on by global warming and climate change.

Pelling's analysis of flood events incorporates a political economy perspective from the urban and regional planning literature to examine the social–environmental dimension of hazardous risks in urban and industrial development (Davis, 1998; Friedmann & Rangan, 1993; Keil et al., 1998). This perspective is applied to a comparison of the political ecology of flood risk in contrasting neighborhoods of the urban area of Georgetown, Guyana. Environmental risk continues to evolve as an important nature–society focal point within political ecology, similar to its ongoing fruitfulness as a focal point in related subfields within geography and other disciplines (Davis, 2001).

Erik Swyngedouw's chapter (Chapter 5) utilizes a political ecology perspective to examine the geographic role of water resource management in Spain's modernization, which centered on urban industrial growth. Swyngedouw develops the concept of "socionature hybrids" to describe the tightly woven linkages between diverse social and political aspects of national modernization and environmental transformations, leading to the creation of

a political–ecological "waterscape" during the period 1890–1930. Methodologically, Swyngedouw extends the qualitative techniques and "cultural turn"
of recent economic geography to a political ecology framework. The design
and methods of Swyngedouw's study involve the textual analysis of hydrological and environmental planning materials as well as broader cultural and historical sources, notably national literary and political documents.

Swyngedouw's study addresses the core themes of this book in two ways.
First, the nature–society theme is directed to the dynamics of environmental
change being linked to Spain's state control and the modern cultural politics of
water resources. The command of nations over resource territories is already a
productive theme in political ecology. Previously this theme of state resource
control has been applied, for the most part, to environmental settings of developing countries as diverse as Southeast Asian forests and coastal areas and African protected areas (Peluso & Watts, 2001; Zerner, 2000). The analysis of water resources in modern Spain in Swyngedouw's chapter demonstrates that the
geographic perspective on state–resource territorialization and the cultural
politics of environmental modernization is of potentially considerable insight
to other countries in Europe and elsewhere.

Second, Swyngedouw's chapter establishes a primary emphasis on modern industry- and urban-related settings as shaping water resources, which he
aptly terms "waterscapes." In political ecology this emphasis can be traced
through a number of interconnecting channels of analysis. Perhaps most notably it builds on studies of modern water development and control, including
relations to agroindustry and urbanization (Emel & Roberts, 1995; Pezzoli,
1998; Swyngedouw, 1997b). These studies offer a rather remarkable demonstration of how the conjoined social–biogeophysical nature of water management lends itself to political ecology analysis. It is striking that various properties of water, including connectivity (flows), transformability (management),
and accountability (measurement), are inherent elements that furnish a social–environmental nature that seems particularly amenable to this theme of
analysis in a geographical approach to political ecology.

PART III. ECOLOGICAL ANALYSIS AND THEORY
IN RESOURCE MANAGEMENT AND CONSERVATION

Political ecology in geography has increasingly sought the incorporation of
ecological analysis. Likewise, more solely environmental analysis—in physical
geography and the environmental sciences—is evermore aware of a political
ecological perspective. The resource systems under analysis are typically
viewed as utilized ecosystems that are, by nature, in ever-changing interaction
with human activities (e.g., people–vegetation, people–wildlife) that are typically differentiated by power relations associated with gender, ethnicity, and
class or wealth category (Bassett & Zimmerer, 2003; Rocheleau, Thomas-Slater,

& Wangari, 1996; Zimmerer & Young, 1998). This perspective of people–environmental interaction builds on a traditional core of geographic studies. Chapters in this section seek to further integrate this core interest into geographical political ecology by focusing on the following: (1) concepts and analysis of geographical scale as a primary factor in the ecological dynamics of resource management; (2) complex human–environment couplings in the change processes that are associated with human impacts and conservation efforts; and (3) multimethod research designs that combine ecological and social scientific analyses.

The chapter by Thomas J. Bassett and Koli Bi Zuéli makes use of plant ecology methods to analyze the relationships of local inhabitants' land use activities with tree and grassland vegetation in the savanna–woodland, agropastoral zones of northern Côte d'Ivoire. An expansion of tree cover is attributed to heightened grazing pressure and a changing fire regime. Their political ecological evaluation of burning contributes a new dimension to policy debates on environmental management of the West African tropics. It rejects the desertification model of advancing aridity along a linear front due to over-cultivation, overgrazing, and deforestation. In conducting this study, the authors relied on a style of integrated methods that is becoming common in political ecological research: (1) ecological analysis (vegetation transects); (2) related image analysis (aerial photo interpretation, geographical information systems [GIS] analysis, ground truthing); and (3) corresponding social–scientific and humanistic approaches (household surveys, oral histories, and discourse analysis).

Bassett and Koli's contribution must be seen amid a quickening stream of sophisticated studies on people–forest interactions in a wide range of settings (Aageson, 1998; Hecht & Cockburn, 1990; Metz, 1994; Moran, 1993). The political economy of globalization, particularly World Bank structural adjustment programs and a spate of neoliberal reforms, such as land privatization, decentralization, and blueprint-based environmental planning, are exerting major impacts on agriculture, livestock raising, and land use (Bassett, 2001). These political economic impacts filter directly into the practices of forest and wildlife access, control, and management in the global South. Consequences include afforestation as well as destruction or degradation of forests (Paulson, 1994; Rocheleau & Ross, 1995; Steinberg, 1998). Multimethod analysis of coupled political–ecological processes is needed to interpret these socio-environmental changes and the challenges they represent. Similar analysis is required to relate vegetation change to the changing political ecology of nontimber forest products (NTFPs), both in developing countries and in the advanced industrial countryside (Hansis, 1998; Voeks, 1998). Coupled dynamics of tree cover and vegetation change, as well as agricultural expansion, must also be further linked to impacts on wildlife populations (Coggins, 2002; Medley, 1998; Naughton-Treves, 1997; Young, 1997).

Karl S. Zimmerer's chapter (Chapter 7) relies on the use of ecological analysis to evaluate the agroecosystems containing world-renowned diverse food plants in rural communities of the Andes Mountains of Peru and Bolivia. Incorporating a perspective on the ecological biogeography of Andean potatoes, maize, quinoa, and ulluco, this chapter evaluates the spatial–environmental patterning of agricultural fields and determines the coupled roles of environmental and social factors. Analysis shows that the predominant pattern consists of "overlapping patchworks," which suggests rethinking the tiered zonation model of mountain agriculture and environmental planning. Overlapping patchworks, rather than layer-cake zonation, offer a variety of specific implications for *in situ* conservation of agrobiodiversity. In undertaking this study Zimmerer made use of multimethod research that combined agroecological sampling (elevation transects of fields), household surveys, analysis of key informative narratives, and cartographic design. Like other studies that employ multimethod research, this chapter uses the different methods to cross-check one another, thus implementing the triangulation technique that is also used by other contributors to this volume (see Bassett & Koli, Chapter 6; Robbins, Chapter 9; and Turner, Chapter 8).

Zimmerer's contribution is situated within related advances on agrobiodiversity and people–mountain studies. Agricultural sustainability and food security are demonstrated to depend on such factors as the access of agriculturalists to the adaptive diversity of seed supplies and the dietary quality of their food plants (Carney, 1996; Cleveland, Bowannie, Eriacho, Laahty, & Perramond, 1995; Grossman, 1998). Government policies, markets for farm products and off-farm labor, and local labor recruiting have impinged on capacities for the cultivation of diverse, nutritionally sound food plants, whose utilization may be primarily possible only for certain sectors of farming populations. Such findings at the people–agriculture interface are leading to calls for programs that support the capacities of farmers to refashion elements of "traditional" and "modern" agriculture into their own amalgams of *in situ* conservation. Gardens have been particularly dynamic sites of political ecological change. In addition, the forum of mountain studies continues to furnish much debate over the nature of agricultural and political–ecological change, with a notable concentration of studies in the Himalayas (Brower & Dennis, 1998; Guthman, 1997; Stevens, 1993).

Matthew Turner's chapter (Chapter 8) employs ecological analysis to build a more realistic appraisal of changes in livestock and rangeland management in the semi-arid environment of Mali in Africa's Sudano-Sahelian region. This ecological analysis shows the importance of shifting spatial and temporal patterns of livestock populations, range resources, and nutrient cycling on rangeland condition. Fluctuations in these ecological properties of rangelands are related in complex ways to herd management. In this chapter, and in his other studies, Turner demonstrates the roles of irregular rainfall, burning, and grazing management in the creation of nonequilibrium range conditions. The herding practices of transhumant pastoralism are, in the main, rational strategies of resource management that are characterized as "opportunistic" since they are highly varied in space and time (Turner, 1999). In carrying out his studies, Turner crafts a combination of research methods that coordinate ecological sampling, nutrient analysis of soils and plant matter, household surveys, historical analysis of market shifts, oral histories with key informants, and GIS and remote sensing analysis.

Turner's work is a major contribution to a building interest in political ecology, namely, the incorporation of nonequilibrium concepts within ecological and human–environmental science (Turner, 1998, 1999). Nonequilibrium ecological dynamics has emerged as a theme that links the social–environmental perspective within political ecology to the broader field of geography (Zimmerer, 1994). It is especially evident in the spatial and temporal distribution of livestock on rangelands that are slated for development and conservation; livestock raising among the Pokot of Kenya, for example, is a "carefully crafted yet flexible 'dance' that is sensitive to time, place, distance, stock-grazing habits, stock endurance, and the happenstance of rain" (Porter & Sheppard, 1998, p. 266). Range degradation arises when the state and global economic forces intrude into rural communities and indigenous institutions of herd management lose their flexibility. Grazing pressure is linked to the quality of herd management and opportunities and constraints affecting herd mobility, particularly herder access to key resources such as high-quality dry season pastures (Bassett, 1994). This series of findings offers a welcome antidote to the conventional concept of grazing pressure as the imbalance between stocking rates and carrying capacity that are spatially and temporally fixed.

PART IV. GEOSPATIAL TECHNOLOGIES AND KNOWLEDGES

Geospatial technologies are increasingly common elements in the methodological toolkit of political ecologists interested in understanding the multiscale dynamics of nature–society relations. The high-tech tools include satellite images, aerial photographs, geographical positioning system (GPS) devices, and GIS mapping and data analysis software. These technologies have most commonly been used to examine regional-scale changes in land use and land cover with the goal of linking these processes to broader scale environmental changes like global warming. This multiscale linking of human and biogeophysical processes is especially true of research undertaken by the human dimensions of the global environmental change community (Turner, 1997; Turner et al., 2001). For political ecologists, global climatic change is just one of many research foci (Byrant & Bailey, 1997; Forsyth, 2001). The breadth of political–ecological studies means that the use of geospatial technologies will vary according to the research questions asked, the scale(s) of inquiry, and data availability (Rindfuss & Stern, 1998). For example, geospatial methods have been effectively used to analyze the dynamics of agricultural intensification and bush encroachment in African savannas (Bassett & Koli, Chapter 6, this volume; Guyer & Lambin, 1993; McCusker & Weiner, Chapter 10, this volume). In general, political ecologists are finding geospatial technologies to be useful in (1) providing empirical data on land use patterns, environmental histories, and biophysical patterns (burn scars, key resources, landscape fragmentation); and (2) linking detailed local studies to broader scales, although the latter is fraught with methodological and epistemological difficulties (Geoghegan et al., 1998; Lambin & Guyer, 1994; Moran & Brondizio, 1998; Turner, 1999; Wiens, 1989). Political ecologists are contributing to the integration of geospatial technologies into social science (Liverman, Moran, Rindfuss, & Stern, 1998), through their explorations of the relationships between geospatial technology, knowledge, and representations of landscapes. The studies in this volume emphasize the productive but contested nature of spatial representation and map making related to the use of remote sensing and GIS techniques.

The chapters by Paul Robbins (Chapter 9) and Brent McCusker and Daniel Weiner (Chapter 10) highlight the methodological and epistemological challenges in integrating geospatial technologies with political ecological research. Both contributions employ these methods in their studies of land use and land cover changes in India and South Africa. Their interest in the political ecological dimensions of landscape change allow them to make a number of important observations about the kind of knowledge produced from using these techniques.

Paul Robbins uses a participatory classification method to compare interpretations of land cover by Indian foresters and agropastoralists in a semi-arid region of Rajasthan. His results show contrasting estimates of the area in forest and wasteland between these two groups, which he links to their divergent evaluations of the utility of the invasive species *Prosopis juliflora*. Foresters view its widespread occurrence as signifying "forest," while agropastoralists classify it as a degraded rangeland due to its low forage potential. This "greening" of the forester's landscape is sanctioned by the use of satellite images and GIS image-processing software that suggest that the forest cover has been objectively measured. Robbins's focus on interpretive struggles between foresters and agropastoralists over land cover and its quality highlights the highly subjective nature of image processing and map making. He argues that these technologies have the effect of actually shaping landscapes since the growth of *juliflora* is encouraged for its reflectance value to foresters, whose bureaucratic careers advance if they can quantitatively demonstrate that forest cover is expanding.

The apparent objectivity of satellite data receives further scrutiny by

McCusker and Weiner, who show how satellite images tend to "naturalize" the landscape by erasing the historical traces of apartheid policies in South Africa on the land. McCusker and Weiner's political ecological approach to geospatial technologies leads them to go beyond the positivist imaging of land cover change to inquire about the social processes producing landscapes. The challenge, they argue, is to uncover these "landscape power relations" that help to *explain* land use and land cover change. Similar to Robbins's participatory mapping methods, they propose a "participatory GIS" in which local knowledge is incorporated into satellite image interpretation and map making. In their hands, geospatial techniques are part of a copious methodological tool kit that also includes survey research techniques, oral histories, group transect walks, and archival research. Used together, these complementary modes of information gathering and analysis can explain land use patterns like agricultural intensification or disintensification in the postapartheid landscape. This merging of geospatial technologies and political ecological investigation through participatory GIS led to the production of land cover maps that sometimes differed from those produced by the authors in their spatial analysis laboratories. Making sense of these discrepancies led to the uncovering of "hidden political ecologies" by which satellite image information revealed the political economic strategy of land users who fabricated a land use history to validate their land claims.

The participatory GIS and mapping methods employed by Robbins and McCusker and Weiner demonstrate that GIS maps are social constructions whose content (e.g., resolution, scale, classification) is the product of innumerable and often conflicting interpretations. Whose categories are accepted and whose interpretation is authorized reflects a politics of selection and omission that is at the center of all map making. The greening of Rajasthan foresters' maps or the erasure of sociospatial processes shaping South Africa's postapartheid landscape in low resolution satellite images reveals as much if not more about the intentions of the map maker and the limitations of the technology than it does about landscape history. These chapters point to the potential gains and limitations of integrating geospatial technologies into political ecological research. Together they advance the "GIS and Society" literature by making more transparent the subjective nature of image interpretation and drawing the links between landscape representations and power relations.

PART V. NORTH–SOUTH ENVIRONMENTAL HISTORIES

Much political ecological research is characterized through the use of a historical perspective; typically, it accounts for the recent past framed at the decades-long scale and, when called for, it commonly incorporates the time scale of colonial precedents. Due to this emphasis, much geography-centered political ecology is fruitfully engaged with the field of environmental history. Indeed, the core of our volume is centered on three areas proposed as touchstones of environmental history: cultural–mental, political–socioeconomic, ecological–biogeophysical. This section of the book especially complements a growing interest in international environmental history, illustrated in work on Mexico (Simonian, 1995). The work here adds a geographical emphasis on the dynamic role of landscapes and human–environmental interaction. These chapters also contribute a series of innovative geographical frameworks that link the histories of environmental and agricultural change, property laws, and ideas of nature in the urban, industrialized countries of the global North with the extensive but less known countryside worldwide.

Andrew Sluyter's chapter (Chapter 11) illuminates the political and ecological origins of the "pristine myth" of a wilderness-type nature that is widely thought to have characterized the New World of present-day Latin America prior to European contact. Sluyter shows how this influential myth, which has powerfully shaped modern environmentalism, arose in the context of Latin America's early colonial period from the combination of environmental changes, such as forest regeneration, demographic decimation of Native American populations, and Spanish colonial ideology that erased the land rights and landscape presence of Native Americans. Sluyter's study helps to bring an international geographical analysis of the pristine myth into a sustained and focused dialogue with environmental history. The latter field has demonstrated the formative influence of the colonial and frontier experiences in the United States and Canada but has devoted far less attention to international legacies. By emphasizing the forming of colonial and postcolonial ideas of pristine nature through conceptual transformations, Sluyter also builds upon an earlier series of pioneering studies in geographical cultural ecology that have highlighted the pre-European impacts of Native Americans (Butzer, 1993; Denevan, 1992, 2000; Doolittle, 1992; Gade, 1992, 1999; Sluyter, 1994).

Roderick P. Neumann's chapter (Chapter 12) illuminates the political, geographical, and cultural dimensions of North–South relations in the establishment of Serengeti National Park, an African protected area that is renowned worldwide. Serengeti was the first national park in Africa. Neumann explains how the park took shape through the interplay of Northern colonialism and the management of Southern environments for the purpose of nature protection. According to Neumann, the British Empire's preconceived, culturally constituted vision of Africa as primeval wilderness and its political economic assumptions of private property became translated into strictly bounded and depopulated territories for wildlife protection in East Africa. Neumann's analysis shows that European policies in East Africa created contrasting yet adjoining landscapes of protected areas and lands of primary economic production that sometimes became severely damaged through agriculture, ranching, and forestry.

One way that Neumann's work is situated within international environmental history is by its reference to the Yellowstone Model that has been a particularly influential blueprint of nature conservation (Hecht & Cockburn, 1990; Stevens, 1997; also see Sundberg, Chapter 3, this volume). The Yellowstone Model is premised on the creation of bounded national parks and the exclusion of people who often formerly inhabited these same areas. Rather than signaling socially progressive attitudes, the designation of reserves under strict protection has been disastrous for many local populations whose livelihoods and human rights have been undermined as a result of reduced access to resources. Contemporary administrators of protected areas have continued in many cases to disregard local claims to resources and territory as well as indigenous environmental knowledge, thus "coercing conservation" (Peluso, 1993). The work of Neumann and others has highlighted how conservation and environmental management frequently invoke issues of human rights. Concerns for human rights often arise through violations, though, in at least a few cases such as those of the Suma and Mosquito people of Honduras and Nicaragua, human rights may be protected proactively (Herlihy, 1993; Nietschmann, 1997).

Judith Carney's chapter (Chapter 13) offers a provocative analysis of the role of African American slaves in the international environmental history of the Americas. It focuses on the introduced agricultural landscapes of domesticated African rice (*Oryza glaberrima*), which stretched from the southeastern United States (particularly the Carolinas), through the Caribbean, and into Brazil and other coastal South American locales. Carney's chapter establishes that it was the technical expertise, cultural traditions, and social strategizing of African American slaves that successfully transplanted a rice landscape into the humid or irrigated lowlands of much of the Americas. Her chapter opens up broad and compelling new vistas on the environmental history of North–South relations that involve groups of people such as slaves that are subordinate within larger societies. Focusing on the 17th and 18th centuries, Carney's chapter is framed in a time period that is intermediate within the context of the other chapters in this section.

Notable in our view is that Carney's work pioneers a geographic and historical framework for diasporic, North–South political ecology (Carney, 2001). This view reveals the role of African cultivators and African American slaves in transcontinental legacies of landscape changes. In doing so, Carney's study creates a historical vision that encompasses the integration of political-social relations of power *and* the dynamics of ecological–biogeophysical processes (i.e., the organisms, technologies, and landscapes of rice growing and processing). Carney's vision of a diasporic political ecology and international environmental history is also connected to advances in a suite of related studies. For example, a more cultural ecological emphasis is evident in the depiction of everyday environmental lifeways under Spanish, Portuguese, and British colonialism (Butzer & Butzer, 1993; Gade, 1992, 1999; Voeks, 1997). These studies reveal the prominence of hybrid cultural ecologies of indigenous, mestizo, creole, and African American peoples. They show that environmental facets of these ecological encounters ranged from biota (crops, weeds, livestock) to tools, technologies, and land use institutions (communal arrangements for management of crops, livestock, and other resources).

CONCLUSION: NATURE–SOCIETY FUSIONS AND SCALES OF INTERACTION

This book is built around the idea of a geographical approach to political ecology as an analytically and methodologically sophisticated framework at the interface of social and natural sciences that differs from more largely social scientific studies as well as more strictly ecological ones. Our volume's focus on this interface raises important new issues of environmental interdisciplinarity (the linking of disciplines) and transdisciplinarity (the fine interweaving of disciplinary analysis), which are addressed more fully in the concluding chapter (Chapter 14). As already mentioned, our approach is *a* geographical political ecology, *one* of several current approaches and a complement to the highly diverse field of inquiry that defines political ecology. These related approaches, each of which is highly promising and equally central as our own, can be distinguished as the following: (1) a primary emphasis on broadly defined environmental politics of social and institutional practices and the "discursive turn" to textual and cultural-political analysis (Braun & Castree, 1998; Bryant, 2001; Bryant & Bailey, 1997; Friedmann & Rangan, 1993; Moore, 1998; Peet & Watts, 1996; Peluso & Watts, 2001; Rangan & Lane, 2001); (2) a primary emphasis on the political economy of nature and commodification in particular (Goldman, 1998; Schroeder, 1995, 1999); (3) a primary emphasis on "politicized" biogeophysical changes and evidence stemming from environmental science and physical geography (Horta, 2000; Stott & Sullivan, 2000); and (4) feminist political ecology with an emphasis on gender analysis (Carney, 1996; Fortmann, Antinori, & Nabane, 1997; Rocheleau, Thomas-Slayter, & Wangari, 1996). It must also be noted that our approach is attuned to other subfields in socio-environmental studies. These broadly similar approaches include cultural and human ecology, environmental sociology, ecological anthropology, household-resource economics, and the types of landscape ecology, agroecology, and conservation biology that have a social science component.

To be sure, the different strands of political ecology are often interwoven to certain extents and collectively they make up a vibrant albeit wide-ranging "field of inquiry" (Peet & Watts, 1996). The broadened scope of policy frameworks for sustainability and environmental risk mitigation, as well as the expanded discussion of philosophical approaches such as critical realism, suggest some of the practical and theoretical ways for integrating the natural and social sciences that can be brought together in political ecology (Blaikie, 1998; Forsyth, 2001; Gandy, 1996; Nyhus, Westley, Lacy, & Miller, 2002). Although the strands of political ecology do not pretend to cement a single or codified theory, they do nonetheless show a concentration of interest and a growing co-alescence of research and policy networks among academics, certain government agencies, and NGOs. The latter include many regional and national organizations whose interests entwine with environment and development issues. Our book is designed with the goal of contributing to this broad interdisciplinarity of political ecology by engaging with issues, contributions, and debates that are evident in the overlap of various sorts of inquiry.

Within the interdisciplinary view, the book is designed with the more specific intent of contributing to political ecology as a vibrant transdisciplinary perspective that fuses the social and biogeophysical sciences. The casting of political ecology as transdisciplinary is as old as the approach and is also newly invigorated (Blaikie, 1994; Blaikie & Brookfield, 1987; Forsyth, 1998; Pezzoli, 1997; Watts, 2000, 2003; Zimmerer, 2000). It is also in a general accord with the views of environmental, conservation, international, and development studies, both within academia and in such realms as governmental and NGO policy (Heberlein, 1988; Vedeld, 1994). Here political ecology benefits from strong tendencies to integrative analysis, close attention to complementarities of styles of reasoning (including the logic of argumentation and evidentiary rules of proof), and openness to varied approaches.

At the same time, *Political Ecology: An Integrative Approach to Geography and Environment–Development Studies* is committed to working along a more specific and particularly productive branch of this overall interdisciplinary-style network of inquiry and analysis. Specifically, our approach seeks to further a theoretically informed perspective that joins the twin geographic themes of nature–society interaction and the political ecology of scale. In doing so, this perspective grows out of our view of the field of political ecology as grounded in intellectual, scientific, and political endeavor. It is our view that the orientations of disciplines continue to serve as home bases for the inter- and transdisciplinary areas that are being established with more frequency in academia and its extensions into the workplace, policy arenas, and project sites. These inter- and transdisciplinary areas are especially apparent in today's world of environmental analysis, witnessed in the many avenues of environmental studies (Carpenter & Turner, 2000; Czech, 2002; Malone & Rayner, 2001).

As a result, this volume looks at the advance and evolution of a geography-centered political ecology within the context of coexisting disciplines. In presenting and shaping the studies here, the book seeks to advance an approach to political ecology that is related to, yet distinct from, other political ecological perspectives. No less important is that our style of political ecology is intended for teaching purposes in order to provide a coherent, recognizable, and pedagogically useful view of this field. We believe the contribution to teaching is vital since the broader spectrum of political ecology may appear diffuse or fragmented to beginning and intermediate-level students.

Indeed, varied interdisciplinary and transdisciplinary approaches to political ecology are multiplying within and among other disciplines, notably sociology (including the sociology of science or "science studies"), anthropology, and planning. Core ideas are also being developed in a still larger range of other disciplines, such as political science, resource economics, conservation biology, and resource management studies. Similarly important to political ecology are advances in such interdisciplinary fields as environmental studies and technology-and-science studies. Our approach complements the developments within these fields and among them. At the same time, it makes a case for the distinctive contributions that this geographic approach is making to the understanding of environmental change, with its core focus on human–environmental interactions and the political ecology of scale. These two foci define our geographical approach to political ecology—one that would tend to be subsumed or made secondary by the different emphasis of those approaches rooted in other disciplines or interdisciplinary areas.

Our volume's dovetailing of these twin themes contributes to the role of political ecology as a key "bridging area" in human–environmental studies and those focused on environmental–development relations in particular (Belsky, 2002; Blaikie, 1998; Forsyth, 1998; Pezzoli, 1998). Our contribution stems from the tight interweaving of social–environmental interactions. But in addition it draws on the relations of the twin themes themselves. Scale integration, spatial as well as temporal, is central to linkages and fusions of social and environmental analysis (Brunckhorst & Rollings, 1999; Fry, 2001). First-order emphasis on the political ecology of scale is thus a cornerstone of our integration. It enables the perspectives of the social and biogeophysical sciences to be woven finely together, rather than coexisting side by side as parallel disciplines. Not least, this volume features works that ply the theory–practice divide. Authors of all the chapters have designed their works in order to participate in research scholarship that is intimately engaged with policy issues. All the authors link their scholarship to policy issues and thus demonstrate the contribution of this political ecology to the major human–environmental debates of the 21st century.

REFERENCES

Aageson, D. L. (1998). On the northern fringe of the South American temperate forest: The history and conservation of the monkey-puzzle tree. *Environmental History*, 3(1), 64–85.

Bassett, T. J. (1994). Hired herders and herd management in Fulani pastoralism (northern Côte d'Ivoire). *Cahiers d'Études Africaines, 133–135*, 147–173.

Bassett, T. J. (2001). *The peasant cotton revolution in West Africa: Côte d'Ivoire, 1880–1995.* Cambridge, UK: Cambridge University Press.

Bassett, T. J., & Zimmerer, K. S. (2003). Cultural and political ecology in the 1990s. In C. Wilmott & G. Gaile, *Geography in America, 2000* (pp. 282–307). Oxford, UK: Oxford University Press.

Belsky, J. M. (2002). Beyond the natural resource and environmental sociology divide: Insights from a transdiciplinary perspective. *Society and Natural Resources, 15*(3), 269–280.

Blaikie, P. (1994). *Political ecology in the 1990s: An evolving view of nature and society* (CASID Series No. 13). East Lansing: Michigan State University.

Blaikie, P. (1998). A review of political ecology: Issues, epistemology, and analytical narratives. *Zeitschrift für Wirtschaftsgeographie, 3–4,* 131–47.

Blaikie, P., & Brookfield, H. (Eds.). (1987). *Land degradation and society.* New York: Methuen.

Brandt, J. (1999). Geography as "landscape ecology." *Geografisk Tidsskrift, 1,* 21–32.

Braun, B., & Castree, N. (Eds.). (1998). *Remaking reality: Nature at the millennium.* London: Routledge.

Brower, B., & Dennis, A. (1998). Grazing the forest, shaping the landscape?: Continuing the debate about forest dynamics in Sagarmatha National Park, Nepal. In K. S. Zimmerer & K. R. Young (Eds.), *Nature's geography: New lessons for conservation in developing countries* (pp. 184–208). Madison: University of Wisconsin Press.

Brunckhorst, D. J., & Rollings, N. M. (1999). Linking ecological and social functions of landscapes: I. Influencing resource governance. *Natural Areas Journal, 19*(1), 57–64.

Bryant, R. (1998). Power, knowledge, and political ecology in the third world: A review. *Progress in Human Geography, 22*(1), 79–94.

Bryant, R. (2001). Politicized moral geographies: Debating biodiversity conservation and ancestral domain. *Political Geography, 19,* 673–705.

Bryant, R., & Bailey, S. (1997). *Third world political ecology.* London: Routledge.

Buttel, F. H. (1987). New directions in environmental sociology. *Annual Review of Sociology, 13,* 465–488.

Buttel, F. H. (2000). Ecological modernization as social theory. *Geoforum, 31,* 57–65.

Butzer, K. W. (1993). No Eden in the New World, *Nature, 362,* 15–17.

Butzer, K. W., & Butzer, E. K. (1993). The sixteenth century environment of the central Mexican Bajío: Archival reconstruction from colonial land grants and the question of Spanish ecological impact. In K. Mathewson (Ed.), *Culture, form, and place: Essays in cultural and historical geography* (Geoscience and Man, Vol. 32, pp. 89–124). Baton Rouge: Louisiana State University.

Carney, J. A. (1996). Rice milling, gender, and slave labor in colonial South Carolina. *Past and Present, 153,* 108–134.

Carney, J. A. (2001). *Black rice: The African origins of rice cultivation in the Americas.* Cambridge, MA: Harvard University Press.

Carpenter, S. W., & Turner, M. (2000). Opening the black boxes: Ecosystem science and economic valuation. *Ecosystems, 3,* 1–3.

Cleveland, D. A., Bowannie, F., Jr., Eriacho, D. F., Laahty, A., & Perramond, E. (1995). Zuni farming and United States government policy: The politics of biological diversity in agriculture. *Agriculture and Human Values, 12*(3), 2–18.

Coggins, C. R. (2002). Ferns and fire: Village subsistence, landscape change, and nature conservation in China's southeast uplands. *Journal of Cultural Geography, 19*(2), 129–159.

Cronon, W. (1991). *Nature's metropolis: Chicago and the Great West.* New York: Norton.

Cronon, W. (1992). A place for stories: Nature, history, and narrative. *Journal of American History, 78,* 1347–1376.

Cronon, W. (1996). The trouble with wilderness; or, Getting back to the wrong nature. In W. Cronon (Ed.), *Uncommon ground: Rethinking the human place in nature* (pp. 69–90). New York: Norton.

Czech, B. (2002). A transdisciplinary approach to conservation land acquisition. *Conservation Biology, 16*(6), 1488–1497.

Daniels, R., & Bassett, T. J. (2002). The spaces of conservation and development around Lake Nakuru National Park, Kenya. *Professional Geographer, 54*(4), 481–490.

Davis, M. (1998). *Ecology of fear.* New York: Metropolitan Books.

Davis, M. (2001). *Late Victorian holocausts: El niño famines and the making of the third world.* New York: Verso.

Delaney, D., & Leitner, H. (1997). The political construction of scale, *Political Geography, 16*(2), 93–97.

Denevan, W. M. (1992). The pristine myth: The landscape of the Americas in 1492. *Annals of the Association of American Geographers, 82*(3), 369–385.

Denevan, W. M. (2000). *Cultivated landscapes of native Amazonia and the Andes.* Oxford, UK: Oxford University Press.

Doolittle, W. (1992). Agriculture in North America on the eve of contact: A reassessment. *Annals of the Association of American Geographers, 82*(3), 386–401.

Eden, S. (2001) Environmental issues: Nature versus the environment? *Progress in Human Geography, 25*(1), 79–85.

Emel, J., & Roberts, R. (1995). Institutional reform and its effect on environmental change: The case of groundwater in the southern high plains. *Annals of the Association of American Geographers, 85*(4), 664–683.

Fairhead, J., & Leach, M. (1996). *Misreading the African landscape: Society and ecology in a forest–savanna mosaic.* Cambridge, UK: Cambridge University Press.

Forsyth, T. (1998). Mountain myths revisited: Integrating natural and environmental science. *Mountain Research and Development, 18*(2), 107–116.

Forsyth, T. (2001). Critical realism and political ecology. In J. Lopez & G. Potter (Eds.), *After postmodernism: An introduction to critical realism* (pp. 146–154). New York: Athlone.

Fortmann, L., Antinori, & Nabane, N. (1997). Fruits of their labors: Gender, property rights, and tree planting in two Zimbabwe villages. *Rural Sociology, 62*(3), 295–314.

Friedmann, J., & Rangan, H. (Eds.). (1993). *In defense of livelihood.* West Hartford, CT: Kumarian Press.

Fry, G. L. A. (2001). Multifunctional landscapes: Towards transdisciplinary research. *Landscape and Urban Planning, 57*(3–4), 159–168.

Gade, D. W. (1992). Landscape, system, and identity in the post-conquest Andes. *Annals of the Association of American Geographers, 82*(3), 460–477.

Gade, D. W. (1999). *Nature and culture convergent: Geography, historical ecology, and ethnobiology in the Andes.* Madison: University of Wisconsin Press.

Gandy, M. (1996). Crumbling land: The postmodernity debate and the analysis of environmental problems. *Progress in Human Geography, 20*(1), 23–40.

Geoghegan, J., Pritchard, L., Jr., Ogneva-Himmelberger, Y., Chowdhury, R. R., Sanderson, S., & Turner, B. L., II. (1998). "Socializing the pixel" and "pixelizing the social" in land-use and land-cover change. In D. Liverman, E. Moran, R. R. Rindfuss, & P. C. Stern (Eds.), *People and pixels: Linking remote sensing and social science* (pp. 51–69). Washington, DC: National Academy Press.

Goldman, M. (Ed.). (1998). *Privatizing nature: Political struggles for the global commons.* New Brunswick, NJ: Rutgers University Press.

Gottlieb, R. (1993). *Forcing the spring.* Washington, D. C. : Island Press.

Grossman, L. S. (1998). *The political ecology of bananas: Contract farming, peasants, and agrarian change in the eastern Caribbean.* Chapel Hill: University of North Carolina Press.

Guthman, J. (1997). Representing crisis: The theory of Himalayan environmental degradation and the project of development in post-Rana Nepal. *Development and Change, 28*(1), 45–69.

Guyer, J. I., & Lambin, E. (1993). Land use in an urban hinterland: Ethnography and remote sensing in the study of African intensification. *American Anthropologist, 95*(3), 17–37.

Hansis, R. (1998). A political ecology of picking: Non-timber forest products in the Pacific Northwest. *Human Ecology, 26*, 49–68.

Heberlein, T. A. (1988). Improving interdisciplinary research: Integrating the social and natural sciences. *Society and Natural Resources, 1*, 5–16.

Hecht, S. B., & Cockburn, A. (1990). *Fate of the forest* (2nd ed.). New York: Harper-Collins.

Herlihy, P. H. (1992). Wildlands conservation in Central America during the 1980s: A geographical perspective. *Yearbook, Conference of Latin Americanist Geographers, 17–18*, 31–43.

Herlihy, P. H. (1993). Securing a homeland: The Tawakha Sumu of Mosquitia's rain forest. In J. Dow & R. V. Kemper (Eds.), *State of the peoples: A global human rights report on societies in danger* (pp. 54–62). Boston: Beacon Press.

Herlihy, P. H. (1997). Indigenous peoples and biosphere reserve conservation in the mosquitia rain forest corridor, Honduras. In S. Stevens (Ed.), *Conservation through cultural survival: Indigenous peoples and protected areas* (pp. 99–129). Washington, DC: Island Press.

Horta, K. (2000). Rainforest: Biodiversity conservation and the political economy of international financial institutions. In P. Stott & S. Sullivan (Eds.), *Political ecology: Science, myth and power* (pp. 179–202). London: Arnold.

Jonas, A. (1994). The scale politics of spatiality. *Environment and Planning D: Society and Space, 12*, 257–264.

Keil, R., Bell, V. J., Penz, P., & Fawcett, L. (Eds.). (1998). *Political ecology: Global and local.* London: Routledge.

Lambin, E. F., & Guyer, J. I. (1994). *The complementarity of remote sensing and anthropology in the study of complex human ecology* (Working Paper 175). Boston University, African Studies Center, Boston, MA.

Lipietz, A. (1995). *Green hopes: The future of political ecology.* Cambridge, UK: Polity Press.

Liverman, D., Moran, E., Rindfuss, R. R. & Stern, P. C. (Eds). (1998). *People and pixels: Linking remote sensing and social science*. Washington, DC: National Academy Press.

Malone, E. L., & Rayner, S. (2001). The role of research standpoint in integrating global- and local-scale research. *Climate Research, 19*, 173–178.

Marston, S. (2000). The social construction of scale. *Progress in Human Geography, 24*(2), 219–242.

McCarthy, J. (2002). First world political ecology: Lessons from the Wise Use Movement. *Environment and Planning A, 34*(7), 1281–1302.

Medley, K. E. (1998). Landscape change and resource conservation along the Tana River, Kenya. In K. S. Zimmerer & K. R. Young (Eds.), *Nature's geography: New lessons for conservation in developing countries* (pp. 39–55). Madison: University of Wisconsin Press.

Metz, J. J. (1994). Forest product use at an upper elevation village in Nepal. *Environmental Management, 18*(3), 371–390.

Moore, D. S. (1998). Clear waters and muddied histories: Environmental history and the politics of community in Zimbabwe's Eastern Highlands. *Journal of South African Studies, 24*(2), 377–403.

Moran, E. (1993). Deforestation and land use in the Brazilian Amazon. *Human Ecology, 21*, 1–21.

Moran, E., & Brondizio, E. (1998). Land-use change after deforestation in Amazonia. In D. Liverman, E. Moran, R. R. Rindfuss, & P. C. Stern (Eds.), *People and pixels: Linking remote sensing and social science* (pp. 94–120). Washington, DC: National Academy Press.

Myers, G. A. (1999). Political ecology and urbanisation: Zanzibar's construction materials industry. *Journal of Modern African Studies, 37*(1), 83–108.

Naughton-Treves, L. (1997). Farming the forest edge: Vulnerable places and people around Kibale National Park, Uganda. *Geographical Review, 87*(1), 27–46.

Naveh, Z. (1991). Some remarks on recent developments in landscape ecology as a transdisciplinary ecological and geographical science. *Landscape Ecology, 5*(2), 65–73.

Naveh, Z. (2000). The total human ecosystem: Integrating ecology and economics. *Bioscience, 50*(4), 357–361.

Neumann, R. (1998). *Imposing wilderness: Struggles over livelihood and nature preservation in Africa*. Berkeley and Los Angeles: University of California Press.

Nietschmann, B. (1973). *Between land and water: The subsistence ecology of the Miskito Indians, eastern Nicaragua*. New York: Seminar Press.

Nietschmann, B. (1997). Protecting indigenous coral reefs and sea territories, Miskito Coast, RAAN, Nicaragua. In S. Stevens (Ed.), *Conservation through cultural survival: Indigenous peoples and protected areas* (pp. 193–224). Washington, DC: Island Press.

Nyhus, P. J., Westley, F. R., Lacy, R. C., & Miller, P. S. (2002). A role for natural resource social science in biodiversity risk assessment. *Society and Natural Resources, 15*, 923–932.

Painter, M., & Durham, W. (Eds.). (1995). *The social causes of environmental destruction in Latin America*. Ann Arbor: University of Michigan Press.

Paulson, D. D. (1994). Understanding tropical deforestation: The case of western Samoa. *Environmental Conservation, 21*(4), 326–332.

Peet, R., & Watts, M. (Eds.). (1996). *Liberation ecologies: Environment, development, social movements.* Routledge: London.

Pelling, M. (1997). What determines vulnerability to floods?: A case study in Guyana. *Environment and Urbanisation, 10,* 469–486.

Peluso, N. (1992). *Rich forests, poor people.* Berkeley and Los Angeles: University of California Press.

Peluso, N. L. (1993). Coercing conservation?: The politics of state resource control. *Global Environmental Change, 3*(2), 199–216.

Peluso, N. L., & Watts, M. (Eds.). (2001). *Violent environments.* Ithaca, NY: Cornell University Press.

Pezzoli, K. (1997). Sustainable development: A transdisciplinary overview of the literature. *Journal of Environmental Planning and Management, 40*(5), 549–574.

Pezzoli, K. (1998). *Human settlements and planning for ecological sustainability: The case of Mexico City.* Cambridge, MA: MIT Press.

Porter, P., & Sheppard, E. (1998). *A world of difference: Society, nature, development.* New York: Guilford Press.

Rangan, H., & Lane, M. B. (2001). Indigenous peoples and forest management: Comparative analysis of institutional approaches in Australia and Asia. *Society and Natural Resources, 14*(2), 145–160.

Rindfuss, R. R., & Stern, P. C. (1998). Linking remote sensing and social science: the need and the challenges. In D. Liverman, E. Moran, R. R. Rindfuss, & P. C. Stern (Eds.), *People and pixels: Linking remote sensing and social science.* (pp. 1–27) Washington, DC: National Academy Press.

Robbins, P. (2002). Obstacles to a first world political ecology: Looking near without looking up. *Environment and Planning A, 34,* 1509–1513.

Robbins P., Polderman, A., & Birkenholtz, T. (2001). Lawns and toxins: An ecology of the city. *Cities, 18*(6), 369–380.

Rocheleau, D., & Ross, L. (1995). Trees as tools, trees as text: Struggles over resources in Zambrana-Chacuey, Dominican Republic. *Antipode, 27,* 407–428.

Rocheleau, D., Thomas-Slayter, B., & Wangari, E. (Eds.). (1996). *Feminist political ecology: Global issues and local experience.* London: Routledge.

Rome, A. (2001). *The bulldozer in the countryside: Suburban sprawl and the rise of American environmentalism.* Cambridge, UK: Cambridge University Press.

Schmink, M., & Wood, C. (1987). The "political ecology" of Amazonia. In P. Little, M. Horowitz, & A. F. Nyerges (Eds.), *Lands at risk in the third world: Local-level perspectives* (pp. 38–47). Boulder, CO: Westview Press.

Schroeder, R. A. (1995). Contradictions along the commodity road to environmental stabilization: Foresting Gambian gardens. *Antipode, 27*(4), 325–342.

Schroeder, R. A. (1999). *Shady practices: Agroforestry and gender politics in the Gambia.* Berkeley and Los Angeles: University of California Press.

Simonian, L. (1995). *Defending the land of the jaguar: A history of conservation in Mexico.* Austin: University of Texas Press.

Slater, C. (2000). Justice for whom?: Contemporary images of Amazonia. In C. Zerner (Ed.), *People, plants and justice: The politics of nature conservation* (pp. 67–82). New York: Columbia University Press.

Sluyter, A. (1994). Intensive wetland agriculture in Mesoamerica: Space, time, and form. *Annals of the Association of American Geographers, 84*(4), 557–584.

Steinberg, M. K. (1998). Political ecology and cultural change: Impacts on swidden-fallow agroforestry practices among the Mopan Maya in southern Belize. *Professional Geographer, 50*(4), 407–417.

Stevens, S. (1993a). Indigenous peoples and protected areas: New approaches to conservation in highland Nepal. In L. Hamilton, D. Bauer, & H. Takeuchi (Eds.), *Parks, peaks, and people* (pp. 73–88). Honolulu, HI: East–West Center.

Stevens, S. (1993b). *Claiming the high ground: Sherpas, subsistence, and environmental change in the highest Himalaya.* Berkeley and Los Angeles: University of California Press.

Stevens, S. (Ed.). (1997) *Conservation through cultural survival: Indigenous peoples and protected areas.* Washington, DC: Island Press.

Stott, P., & Sullivan, S. (2000). *Political ecology: Science, myth and power.* London: Arnold.

Swyngedouw, E. (1997a). Neither global nor local: "Glocalization" and the politics of scale. In K. R. Cox (Ed.), *Spaces of globalization: Reasserting the power of the local* (pp. 137–166). New York: Guilford Press.

Swyngedouw, E. (1997b). Power, nature and city: Water and the political ecology of urbanization in Guayaquil, Ecuador: 1880–1990. *Environment and Planning A, 29,* 311–332.

Turner, B. L., II. (1997). The sustainability principle in global agendas: Implications for understanding land-use/cover change. *Geographical Journal, 163*(2), 133–140.

Turner, B. L., II, Cortina Villar, S., Foster, D., Geoghegan, J., Keys, E., Klepeis, P., et al. (2001). Deforestation in the southern Yucatán peninsular region: An integrative approach. *Forest Ecology and Management, 154,* 353–370.

Turner, M. (1998). The interaction of grazing history with rainfall and its influence on annual rangeland dynamics in the Sahel. In K. S. Zimmerer & K. R. Young (Eds.), *Nature's geography: New lessons for conservation in developing countries* (pp. 237–261). Madison: University of Wisconsin Press.

Turner, M. (1999). Merging local and regional analyses of land-use change: The case of livestock in the Sahel. *Annals of the Association of American Geographers, 89*(2), 191–219.

Vedeld, P. O. (1994). The environment and interdisciplinarity. *Ecological Economics, 10,* 1–13.

Voeks, R. A. (1997). *Sacred leaves of Candomblé: African magic, medicine, and religion in Brazil.* Austin: University of Texas Press.

Voeks, R. A. (1998). Ethnobotanical knowledge and environmental risk: Foragers and farmers in northern Borneo. In K. S. Zimmerer & K. R. Young (Eds.), *Nature's geography: New lessons for conservation in developing countries* (pp. 307–326). Madison: University of Wisconsin Press.

Walker, P. A. (2003). Reconsidering regional political ecologies: Toward a political ecology of the rural American West. *Progress in Human Geography, 27*(1), 7–24.

Watts, M. (2003). *For political ecology.* Unpublished manuscript, University of California, Berkeley.

Watts, M. (2000). Political ecology. In E. Sheppard & T. J. Barnes (Eds.), *A companion to economic geography* (pp. 257–274). Malden, MA: Blackwell.

Wiens, J. A. (1989). Spatial scaling in ecology. *Functional Ecology, 3,* 385–397.

Young, K. (1997). Wildlife conservation in the cultural landscapes of the central Andes. *Landscape and Urban Planning, 38,* 137–147.

Zerner, C. (Ed.). (2000). *People, plants, and justice: The politics of nature conservation.* New York: Columbia University Press.

Zimmerer, K. S. (1994). Human geography and the "new ecology": The prospect and promise of integration. *Annals of the Association of American Geographers, 84,* 108–125.

Zimmerer, K. S. (2000). The reworking of conservation geographies: Nonequilibrium landscapes and nature–society hybrids. *Annals of the Association of American Geographers, 90*(2), 356–69.

Zimmerer, K. S., & Young, K. R. (1998). Introduction: The geographical nature of landscape change. In K. S. Zimmerer & K. R. Young (Eds.), *Nature's geography: New lessons for conservation in developing countries* (pp. 3–34). Madison: University of Wisconsin Press.

PART I

PROTECTED AREAS AND CONSERVATION

CHAPTER 2

Balancing Conservation with Development in Marine-Dependent Communities

Is Ecotourism an Empty Promise?

Emily H. Young

Until recently, the bounty of the sea was considered limitless. There is growing evidence, however, of a crisis in the world's fisheries. As commercial fish stocks rapidly become depleted, it is increasingly apparent that current patterns of marine resource use and distribution cannot be sustained (Pauly et al., 2002). Small-scale fishers, who comprise approximately 94% of the world's fishers and catch nearly half of the global fish supply for human consumption, have been particularly hard hit by this crisis (McGoodwin, 1990, p. 8). Consequently, the question of how best to promote greater stewardship of marine resources cannot be resolved without also addressing the needs and interests of the small-scale fishers who depend on such resources for their livelihoods.

There is growing concern over how to balance the conservation of natural resources with development needs worldwide. In Latin America and other developing regions, environmental degradation is often related to problems of unequal access to natural resources and poverty (Little & Horowitz, 1987).

Adapted by Emily H. Young from the article "Balancing Conservation with Development in Small-Scale Fisheries: Is Ecotourism an Empty Promise," published in *Human Ecology*, 27(4), 581–620 (1999). Adapted by permission of Kluwer Academic/Plenum Publishers. The adapting author is solely responsible for all changes in substance, context, and emphasis.

29

Consequently, it is now widely recognized that efforts to link nature conservation to economic development in these regions must address interconnected ecological, resource use, and socioeconomic concerns; moreover, local involvement is often viewed as crucial (Western & Wright, 1994).

Interest in ecotourism as a means to encourage community-based conservation and development is increasingly widespread (Cater & Lowman, 1994). *Ecotourism* is defined by the Ecotourism Society as travel to natural areas that minimizes ecosystem impacts and provides local people with a financially vested stake in conservation (Wood, 1991). It is one of the fastest growing sectors of the global tourism industry (Mowforth & Munt, 1998, p. 97). Indeed, some natural attractions, such as whale watching, have spawned the development of multi-million-dollar tourism enterprises (Hoyt, 1995).

A number of studies have examined the potential for ecotourism to encourage environmentally sound development, generate revenues for protected areas and resident peoples, and provide a nonconsumptive alternative to consumptive resource use (e.g., Cater & Lowman, 1994; Miller & Malek-Zadeh, 1996). However, preliminary assessments of the industry in a variety of cultural and biophysical settings worldwide reveal numerous drawbacks to ecotourism development. These drawbacks include adverse impacts on wildlife and fragile ecosystems from nature-based tourism development (Savage, 1993), the breakdown of local cultural traditions (Brower, 1990), few economic benefits to local people (Place, 1995) or protected areas (Lindberg, Enriquez, & Sproule, 1996), and aggravated conflicts over resource access (Barkin, 1996).

Although the shortcomings of ecotourism have been generally recognized, it continues to be widely promoted as a means to link development with conservation. Whale tourism is heralded as a way to promote conservation in marine environments (Barstow, 1986), yet few comprehensive studies have been conducted to critically evaluate whether it is an economically and socially viable activity for local communities or whether it is compatible with the protection of marine habitats and wildlife (Dedina & Young, 1995). Ultimately, as Brandon and Margoulis (1996, p. 35) point out, the question of whether employment in ecotourism provides sufficient incentive for local people to safeguard natural resources can only be addressed on a site-specific basis.

This study looks at the case of recreational whale watching in the small-scale fishing communities of Laguna San Ignacio and Bahia Magdalena in Baja California Sur, Mexico, to examine whether the economic benefits of ecotourism reduce resource conflicts and extractive pressures on inshore fisheries and promote stewardship of marine resources by local people. With more than 2,200 kilometers, or 23% of the nation's coastline, and over 2,000 marine species, the northwestern state of Baja California Sur lies in the heart of Mexico's prime fishing region (Villareal G., 1987, p. 59). Since the 1970s, inshore commercial fisheries have experienced explosive growth in the region, particularly among small-scale producers. Within a relatively short period, access rights to local fishing grounds have become increasingly contested and some local shellfish and fish stocks have been depleted to the point of commercial extinction (Young, 2001).

Concurrent with the demise of key regional fisheries, tourism has grown rapidly, to become the state's third most important industry, after fishing and mining (INEGI, 1995). Over the last decade, local fishers, governmental planners, and environmental organizations have become increasingly interested in ecotourism through recreational whale watching, touting it as a way to simultaneously pursue the goals of economic development and marine resource conservation by creating an alternative form of local income generation to fishing through nonconsumptive use of local wildlife. To assess whether ecotourism attenuates destructive exploitation of inshore fisheries and promotes conservation, I examine how the inhabitants of both Laguna San Ignacio and Bahia Magdalena cope with conflicts over access to common-pool resources in recreational whale watching.

APPROACH AND METHODS

Political ecology, defined by Blaikie and Brookfield (1987, p. 17) as a multiscalar approach that integrates the ecology and political economy perspectives, has been widely employed to understand the "underlying contexts and processes" (Bassett, 1988, p. 453) of environmental degradation in relation to problems of conservation and the development of land-based resources in a variety of biophysical and cultural settings worldwide. While early studies focused on how global capitalism and state policies shape nature–society relations at various scales in lesser developed regions (e.g., Hecht & Cockburn, 1989), more recent studies have given greater attention to how political processes and power inequalities influence these relations (e.g., Peet & Watts, 1996).

A recurring theme in the political ecology literature addresses conflicts over land, flora and fauna, soil, and water (Bryant, 1992), especially those surrounding common-pool resources (Carney, 1993; Sheridan, 1988). A growing number of studies use a political ecology approach to examine the relationship between access conflicts in the commons and ecological change in aquatic habitats and wildlife, particularly in marine environments (e.g., Stonich, 1995; Young, 2001). This study uses a political ecology approach to examine how area inhabitants use common-pool marine resources for gray whale tourism, as shaped by community-based management practices, regional economic and political structures, and governmental policies.

Information for this study was obtained primarily from 66 oral interviews (33 in each study site) and participant observation of local fishing and tourism activities conducted in Laguna San Ignacio and Bahia Magdalena during 1993 and 1994, as well as during follow-up visits in 1995 and 1997. Because very few published data exist concerning fisheries and ecotourism in Baja California Sur, interviews with area residents provided a critical source of information concerning the growth of these two sectors and for understanding how local people have responded to markets, centralized resource management structures, and policies for marine resource development and conservation. Informal, semistructured interviews were the most conducive method for gathering information from people in both study sites. Local inhabitants were suspicious of outside researchers using standardized questionnaires and tape recorders. They tended to be much more open and communicative in conversation, which could be directed into particular areas of my research. I used participant observation of fishing and tourism activities to corroborate interview data and determine the degree to which attitudes and ideas expressed during interviews were reflected in human activities in surrounding coastal and marine habitats.

I complemented interview data from the field sites with that obtained from nonlocal sources, including (1) 25 municipal, state, and federal officials based elsewhere in Baja California Sur and in Mexico City who were charged with regulating regional fishing and tourism activities; (2) three executives from the largest fish-processing industries that buy fish and shellfish from the study sites; and (3) 21 representatives from tourism enterprises based elsewhere in Mexico and the United States. These interviews provided me with broader insights into the role of governmental policies in marine development and conservation in the study sites and the relationships between local producers/service providers with international and domestic markets for fishing and tourism. I also used primary data in state and federal socioeconomic censuses to obtain information about the local economic activities and population characteristics of each area and to make calculations of changes during the past several decades, which were compared with local inhabitants' estimations of environmental change. Finally, I examined newspaper coverage of environmental issues in the major daily circulated throughout the state of Baja California Sur, the *Sudcaliforniano*, during four gray whale tourism seasons in January–March 1992–1995, and during follow-up research in 1997 and 1998, to verify events, dates, and times mentioned during interviews and to gain a sense of how local marine conservation issues are publicized.

FISHERIES AND GRAY WHALES IN LAGUNA SAN IGNACIO AND BAHIA MAGDALENA

Nearshore waters in and around Laguna San Ignacio and Bahia Magdalena serve as important nurseries for numerous commercially valuable species of fish and shellfish, including California halibut (*Paralichthys californicus*), shortfin corvina (*Cynoscion parvipinnis*), abalone (*Haliotus* spp.), lobster (*Panulirus* spp.), and Pacific calico scallops (*Argopecten circularis*), from which hundreds of small-scale fishing families residing in these areas derive a living. The shallow, calm, and warm waters of these coastal embayments also serve as habitat for gray whales (*Eschrichtius robustus*) which mate, give birth, and rear their infants between December and April, after which time they migrate to summer feeding grounds in the Bering, Beaufort, and Chuckchi Seas.

These enclosed lagoons were favored hunting grounds for European and U.S. whalers who harvested these cetaceans to near extinction from the middle of the 19th to the early 20th centuries (Henderson, 1972). California gray whales have since rebounded to their estimated preexploitation stock size due to various protective measures, including a global ban on their commercial harvest. The once-favored hunting grounds of whalers have increasingly become popular tourist destinations for recreational whale watching during the last three decades (Dedina & Young, 1995) (Figure 2.1).

Laguna San Ignacio is located 700 kilometers south of the United States-Mexico border and forms part of the Vizcaino Biosphere Reserve, established in 1988. This 2.5-million hectare reserve is one of the largest formally protected areas in the Americas. Area inhabitants live in seven settlements scattered around the shores of the lagoon. Relatively isolated, Laguna San Ignacio is 60 kilometers from San Ignacio, the nearest town. The lagoon is linked to San Ignacio by a dirt road. There are no basic services such as running water, sewerage, or electricity at the lagoon. All basic supplies are trucked into the area.

Bahia Magdalena is located some 200 kilometers south of Laguna San Ignacio. In contrast to Laguna San Ignacio, Bahia Magdalena benefits from a well-developed network of roads and basic services. This infrastructural development was an integral part of government-sponsored programs for agricultural colonization of the adjacent Santo Domingo Valley after World War II that were intended to reduce pressures for land elsewhere in Mexico (Barrett, 1974). Bahia Magdalena also harbors three fish canneries, a port for oil tankers and large fishing boats, and a newly constructed power plant. In Bahia Magdalena, the fishing town of Puerto Adolfo Lopez Mateos is the focus of this study. Town residents derive a living from fishing and fish processing in the bay.

Fishing communities in both Laguna San Ignacio and Bahia Magdalena are relatively new and highly heterogeneous. Indigenous populations that once exploited the nearshore waters were virtually wiped out by disease as a result of European contact and colonization (Aschmann, 1959). Pioneer settlers initially migrated from inland ranches and towns to the Pacific Coast on a seasonal basis during the 1920s and '30s to harvest lobster, shark, and sea turtles for sale to merchant ships from San Diego. Given recurrent problems of drought in the region, settlers also fished the waters to supplement their often

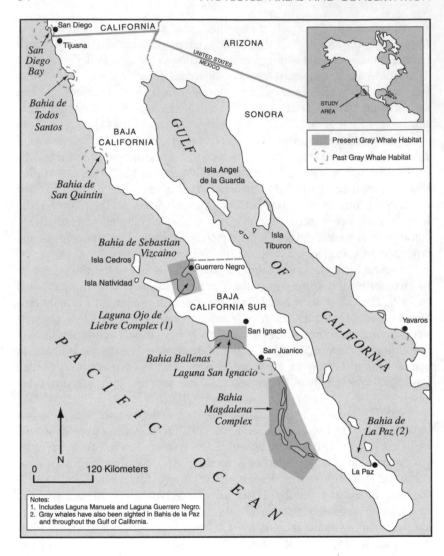

FIGURE 2.1. Baja California peninsula with past and present gray whale habitat and adjacent coastal lands outlined.

meager ranch supplies of cheese, fruit and vegetables, and occasional red meat. Seasonal fish camps gradually grew into permanent settlements as more and more people moved to Laguna San Ignacio and Bahia Magdalena in the hopes of securing a better life by fishing from the sea.

Since World War II, harvesting of fish and shellfish in both Laguna San Ignacio and Bahia Magdalena has increased rapidly with the introduction of new fishing technologies and the growth in commercial fisheries. Similar to the experience of small-scale fishing communities elsewhere (e.g., Nietschmann, 1973; Johannes, 1981), the introduction of gill nets, refrigeration, and motor-powered skiffs has enabled more intensive, commercial harvesting of marine resources. The rapidly growing regional fishing industry attracted unprecedented numbers of immigrants from impoverished rural areas of mainland Mexico seeking greater economic opportunities in fishing and occasionally fleeing political and drug-related violence in their home communities.

Available data on the fishing sector in Baja California Sur for the period between 1980 and 1992 reveal growth in the fishing population by over 400%, from 2,497 to 11,396 fishers (Estado de Baja California Sur, 1987, p. 145; INEGI, 1993). Local population data for both Laguna San Ignacio and Bahia Magdalena reflect these demographic shifts. Between 1970 and 1995, the resident population of Laguna San Ignacio increased by nearly 2,000%, from 26 to 502. In Bahia Magdalena, the local population rose by 227%, from 3,043 to 6,930, during the same period. The town of Puerto Lopez Mateos absorbed over one-third of all newcomers to the bay, growing from a population of 1,283 to one of 2,391 in those years (Direccion General de Estadistica, 1972; INEGI, 1995).

The increase in local fishing populations brought about unprecedented harvesting of nearshore resources (Young, 2001). The staggering growth in statewide fishery production obscured local patterns of overharvesting where some species, such as Pacific calico scallops, became commercially extinct (Macda-Martinez, 1990, p. 10). Although production levels have since fallen for a number of fish and shellfish around the Baja California peninsula, the long-term implications for the region's fisheries are not well understood, as few studies have been conducted on the population dynamics of the various marine species that are commercially exploited in the region. But the rising number of fishers exploiting nearshore waters in both of these areas has created often bitter conflicts among multiple users over access to marine resources that are becoming increasingly scarce.

Concurrent with the demise of regional fisheries has been the growth in recreational whale watching along the Baja California peninsula. From the early 1970s onward, tourism companies based in the United States have brought rising numbers of foreign visitors on package tours to Laguna San Ignacio and Bahia Magdalena to see gray whales. Few local people worked in this industry, however, until recently. Since the late 1980s, as growing numbers of tourists have traveled on their own to these areas to catch a glimpse of the marine mammals at firsthand, more and more fishers in both areas have begun to hire out skiffs and serve as tour guides for recreational whale watching. In the face of deteriorating area fisheries, gray whale tourism is increasingly viewed as a promising source of income during the low fishing season between December and April.

I next examine the environmental conditions, socioeconomic and political structures, and governmental policies that have undermined the ability of small-scale producers to cope with the common-pool resource status of coastal marine areas, fueling conflicts over marine resources and undermining incentives for resource stewardship.

COMMON-POOL RESOURCE CONFLICTS IN MARINE ENVIRONMENTS

Along the Pacific Coast of the Baja California peninsula, small-scale producers have increasingly become embroiled in conflicts over access to common-pool resources in coastal marine environments. Most aquatic fauna are by nature highly mobile. Consequently, control over access to and use of aquatic resources is inherently problematic. They are typically considered to be common-pool or common-property resources, which implies a class of goods for which it is difficult to exclude users and the goods exploited by one user subtract from those available to others (McCay & Acheson, 1987).

Garret Hardin's (1968) "The Tragedy of the Commons" popularized the idea that resources held in common (e.g., fish, forests, pasture) are inevitably abused in a way that privately or publicly owned resources are not (Ciriacy-Wantrup & Bishop, 1975). The tragedy of the commons has often been used to explain why marine resources are frequently overexploited (Berkes, 1985). Underlying Hardin's explanation are three key assumptions: (1) common property rights are equivalent to open access; (2) people are basically selfish; and (3) common-pool resources are vulnerable to overuse and degradation (Berkes, 1987).

There are various examples, however, of collective arrangements to govern common property rights in a manner that promotes resource stewardship and long-term productivity in coastal marine environments (e.g., Acheson, 1988). McCay (1978) notes that community-based arrangements to govern common-pool resources usually center on regulating access to marine territories. But commercialization and increasingly centralized management of coastal marine resources has often marginalized small-scale producers by eroding community-based access rights and transferring power in the marketplace from producers to secondary processors and retailers, thereby fueling conflicts over resource access as well as destructive resource use (Crean & Symes, 1996, pp. 8–9; Sinclair, 1990).

Because their migratory range extends over 5,000 miles, gray whales are not the exclusive domain of any one group, but are instead exploited by multiple users operating independently of one another throughout their habitat range. In their Arctic feeding grounds, for instance, they are still occasionally harvested for subsistence use by Siberian Natives in the Russian Federation. During their southward migration along the eastern Pacific coast and their stay in the breeding and calving lagoons, they are exploited for nonconsumptive purposes by the tourism industry. Because gray whales tend to use nearshore habitats for feeding, breeding, and migrating, they are vulnerable to direct physical harm and habitat degradation from numerous other human activities (e.g., commercial fishing, merchant shipping, offshore oil and gas exploration, and coastal development) (Marine Mammal Commission, 1994).

Until the early to mid-1990s, recreational whale watching in Laguna San Ignacio and Bahia Magdalena was officially open to any domestic or foreign tourism enterprise that met minimum government requirements. Consequently, the million-dollar industry has been dominated by United States-based operators who have greater access to sufficient capital to invest in developing tour facilities and marketing. With the growing popularity of this activity, a greater number of area residents have started hiring out their services to tourists, primarily as guides. As the tourism industry has grown, so have the conflicts among local and outside-based operators over the issue of access rights to the water, whales, and tourists.

COLLECTIVE INABILITY TO MITIGATE RESOURCE CONFLICTS IN MARINE ECOTOURISM

Community-Based Resource Management

In the past, a poorly developed tourism infrastructure (in terms of local lodging facilities, restaurants, transit for tourists, and basic services[1] such as electricity, running water, and sewerage) has served as a de facto mechanism for limiting the growth of tourism and minimizing tourist impacts on local habitat and wildlife. During the last decade, however, gray whale tourism has grown rapidly, particularly in Bahia Magdalena, where the municipal planning office estimated that in one year alone—between 1993 and 1994—the number of tourists who came to the bay to see whales rose by 300% (Municipio de Comondu, 1994).

Tour boats are self-contained and tour camps are currently set up on a temporary basis. But if gray whale tourism continues to grow in its present form, problems with inadequate sewage and trash disposal are likely to pose risks to public health and present other environmental hazards. Conversely, attempts to install basic services in Laguna San Ignacio and improve them in Bahia Magdalena to prevent such problems may encourage the expansion of tourist facilities and residential development, which could place an increasing strain on area natural resources and threaten the long-term ecological vitality of coastal marine habitat. At present, there is no consensus among marine biologists on appropriate levels of boating activities for the breeding of gray whales and little is known about the long-term impacts of recreational whale watching from a short distance on these marine mammals (e.g., Jones & Swartz, 1984; Romero G., 1990).

Given its high accessibility, Bahia Magdalena is already experiencing problems. A paved road leading from Highway 1, a major highway, directly to Puerto Adolfo Lopez Mateos makes it possible to reach the town by private passenger car or recreational vehicle, bus, or taxi. An airstrip for private planes and small airlines is located adjacent to the docks for whale observation skiffs. Basic services have been inadequate to absorb the rapidly growing gray whale tourism industry in the area. The town's sewage pipes currently discharge untreated effluent directly into the bay. During the 1994 season, the beach where guided boat tours are offered had only two public latrines and one 50-gallon trash barrel. Environmental degradation from such uncontrolled tourism activities is likely to adversely affect coastal and marine resources, heightening the level of uncertainty faced by local tour guides and fishers to an even greater extent than in Laguna San Ignacio. The rapid growth of such activities also elevates the potential for conflict over access to the water, whales, and tourists.

Many local residents acknowledge that, until recently, they had little interest in gray whales. But as they have come to increasingly appreciate the economic potential of these marine mammals, they have acquired a more personal stake in their long-term survival. Such changes in attitude could be mobilized into a local constituency base for the protection of gray whales in both Laguna San Ignacio and Bahia Magdalena. There are few institutional mechanisms at the community level, however, to reinforce such attitudes and convert them into collective action that could ensure the long-term survival of gray whales and the ecological vitality of their habitat. Because this problem plays out differently in each place, the cases of Laguna San Ignacio and Bahia Magdalena are best treated separately.

Laguna San Ignacio

In Laguna San Ignacio, the local residents who hire out skiffs for recreational whale watching operate more as fiefdoms than as collective ventures. Two main families form the basis of each fiefdom. Much of the competitive tension between fiefdoms, however palpable, remains beneath the surface. During the 1994 season and follow-up visits in 1995 and 1997, I noted few outward manifestations of competition other than mutual bad-mouthing and the rapid proliferation of signs (and countersigns) advertising whale guide services along the road running into and around the lagoon.

A greater source of local conflict has been the structural changes in one of the area's *ejidos* (communally held lands with certain legal entitlements) in re-sponse to recent changes in Article 27 of the Mexican Constitution that permit the formation of private enterprises and the sale of communal lands. During an *ejido* meeting in December 1992, the board of directors, all of whom have immigrated to the area from other parts of Mexico, initiated proceedings to form a private gray whale tourism company. By January 1994, they had erected a small *palapa*-style restaurant, a few outhouses, and a handful of tents to ac-commodate individual visitors and organized tour groups for recreational whale watching. By using its *ejido* affiliation to apply for whale-watching per-mits, the company was able to obtain exclusive rights to service foreign tour boats and camps with guided skiffs. Other company employees were from the town of San Ignacio or another town further north, Guerrero Negro. The com-pany only employed one lagoon inhabitant as a skiff driver in 1994. This fact was not lost on other area residents, many of whom complained that few local people were given the opportunity to work in gray whale tourism. Perceived injustices in the distribution of benefits from tourism also cultivated senti-ments of ill-will among some area residents toward the *ejido*, as captured by this fishers' comments: "The *ejido* pays [bribes] so that they get preference over others for permits to take out tourists. The *ejido* doesn't serve [anyone] here. . . . " Since the 1993–1994 season, the company has hired more lagoon residents as skiff drivers. Whether these hiring efforts will help to alleviate ten-sions surrounding *ejido* efforts remains uncertain, given other conflicts that have emerged surrounding land tenure.

Currently, all beachfront lands are considered prime for gray whale tour-ism activities. The directors of the *ejido* recently parceled out *ejido* lands for private holdings among its members. *Ejido* directors contend that the meet-ings in which the formation of the private company and the division and dis-tribution of *ejido* lands were carried out were publicized and open to all *ejido* members. Other *ejido* members, however, including a family whose house now sits on land that was parceled out to someone else, claim that they were never informed of the meeting. Other long time lagoon residents who never joined the *ejido* have similar complaints. These land tenure disputes signal the accel-eration of local conflicts with the growth of gray whale tourism in the lagoon. Unless all parties involved take steps toward reconciliation and resolution of these problems, this situation could deteriorate in the future.

Bahia Magdalena

Whether as members of labor unions in the factory or of fishing cooperatives at sea, the residents of the town of Puerto Adolfo Lopez Mateos have been able to more effectively organize to promote local interests. This is most evident in the formation of the local tourism cooperative. Members who were inter-viewed asserted that forming a cooperative was the most viable option open to those who wanted to work in gray whale tourism but had few financial resources and little experience in running a private business. They also noted that cooperatives served as a vehicle to enable a larger number of town residents to work as whale skiff guides than might otherwise be possible and to rotate access to tourists among its members to ensure a more equitable distribution of tourism benefits.

While the cooperative arrangement has proven beneficial, it has not completely eliminated conflicts over access rights within the community. There are complaints from within and outside the cooperative about certain cooperative members who have abused their membership by renting out their privileges as whale skiff drivers to others. Others charge that cooperative directors mismanage and occasionally steal from the association's funds. As one cooperative member put it: "I tell you, some people [in the cooperative] are selfish. The president of the cooperative is managing it as if he were the owner. [People who rent out their privileges to others] should give those privileges away to other families who really need the money. . . . " Actions taken by one cooperative member to form a private recreational whale-watching business a few years ago precipitated a family feud and sparked charges by other town residents that he had grown rich overnight at the expense of the town.

During the 1994 season, the only community-based institution that served to regulate local recreational whale-watching activities was the tourism cooperative. Original members of the tourism cooperative note that its founding principles were to regulate boating activities and to minimize competitive conflict between local boat guides. This organization did rotate all local guides through a list that systematically assigned guides to take tourists out to watch whales. The cooperative also set both the price and the maximum tourist capacity per boat at relatively low levels, compared with other recreational whale-watching destinations.[2] Although setting a low price may have given one a competitive advantage vis-à-vis whale tourism elsewhere, it also pressured the cooperative to keep as many boats operating in the water as possible, in order to ensure all of its members an income that was comparable to, if not better than, that earned from fishing.

When another group of 31 aspiring whale skiff guides began to take shape during the course of the 1994 season, tensions over access to gray whale tourism flared between cooperative members and the new group. The formation of the new group challenged the capacity of the cooperative to manage local recreational whale-watching activities. Unless current disagreements over access rights are resolved, the prospects for greater conflict between the two groups are high.

There has been little communication or coordination of activities between the cooperative, the new group, and the numerous outside-based tour camps and boats. The cooperative wields little power in regulating the activities of the latter. Particularly toward the end of the 1994 season, the number of gray whales inside the bay began to dwindle as many started their journey northward toward summer feeding grounds. In order not to disappoint tourists, local and outside-based tour guides would competitively jockey for positions around the few remaining whales in the bay, where it was not uncommon to see a dozen boats around one whale.

In both Laguna San Ignacio and Bahia Magdalena, competitive conflicts over access to marine resources have only intensified with the growth of recreational whale watching. This tends to promote collective behavior in actions geared toward short-term individual interests. Such behavior tends to foster more reckless conduct in tour boats around whales, elevating the risk of injury or death from this activity for both people and whales. Many residents of both areas have expressed concerns that rapid and uncontrolled growth of tour boats around whales may drive these cetaceans to seek refuge elsewhere, and tourism will consequently be doomed to a fate similar to that of area fisheries.

Regional Political Economy

At the regional level, economic and political structures and government policies tend to reinforce—and even exacerbate—conflicts over access to common property resources at the community level. As is the case with ecotourism elsewhere, recreational whale watching is subject to boom-and-bust cycles according to global market and political conditions that are beyond local control (Mowforth & Munt, 1998). Visitors to Laguna San Ignacio and Bahia Magdalena are seasonal. This means that income from ecotourism in seasonal at best, and likely to vary with changing tourist preferences in natural attractions. Although ecotourism provides a new source of income to some area residents, they still rely upon fishing during the non-tourist season.

While recreational whale watching has grown into a multi-million-dollar industry along the Baja California peninsula, a small proportion of tourism revenues remains with the communities involved. Indeed, the economic impact of gray whale tourism is much more significant at the regional and national levels. In both Laguna San Ignacio and Bahia Magdalena, tour boat and whale camp operators based outside of these areas control a large share of the recreational whale-watching market. There are few local restaurants, shops, hotels, or other local businesses that cater to tourists in either area. Consequently, tourists who come to these areas to see whales spend little of their money locally.

In 1994, approximately $3.3 million was spent by tourists visiting the Laguna San Ignacio through package whale-watching tours organized by outside-based companies (Table 2.1). An estimated $40,300 (1.2%) of these revenues was spent on salaries and supplies purchased at the lagoon.[3] By comparison, the three fishing families and one individual from the lagoon who work in tourism do so on a much smaller scale, providing guided boat tours, camping facilities, and home-cooked meals to a growing number of visitors who come to the area on their own to see the whales. During the 1994 season, these local enterprises netted between $2,000 and $6,000 (discounting the costs of gas, oil, skiff maintenance, etc.) from recreational whale watching, comprising approximately one-fourth to one-half of their yearly incomes.

In Bahia Magdalena, of the approximately $5 million that outside-based tourism companies grossed from package whale tours during the same year, outside-based tourism companies spent about $33,000 (< 1%) on local salaries, restaurants, and supplies (Table 2.2). The 44 local residents who were employed in tourism as boat drivers reported net average earnings of $1,400/person, or up to one-half of their total yearly income.

A much higher portion of outside-based tourism companies' revenues is spent on direct operating costs in Mexico outside of Laguna San Ignacio and Bahia Magdalena. Some companies estimated that such costs ran between 50% and 65% of their gross income from recreational whale watching. For instance, land-based tourism companies spent approximately $94,000 in the town of San Ignacio on restaurants, supplies, dining, lodging, and taxi services. In San Carlos and Ensenada, cruise ships and tour boats spent an estimated $87,000 on port costs prior to tours of Laguna San Ignacio.

The Mexican government has taken some steps to remedy this situation. For instance, in 1990, in Laguna San Ignacio, it established a requirement that foreign tour companies hire Mexican skiff drivers who have formal permits for recreational whale watching. The new government requirements, coupled with

TABLE 2.1. Gross Earnings of Outside-Based Tourism Operators in Laguna San Ignacio

Type of tour operation	Number of tourists	Typical length of tour (in days)	Typical period of stay in lagoon (in days)	Average price/ person (in $U.S.)	Tour operators gross earnings (in $U.S.)[a]
Cruise ships (2)	560	8	1.5	3,500	1,960,000
Tour boats (6)	500	8	3	1,500	750,000
High-price tour camps	360	5	3	1,300	468,000
Low-price tour camps	175	3	2.5	800	140,000
TOTAL	1,595				3,318,000

[a]Some companies estimated their direct operating costs in Mexico to be 50–65% of their gross earnings for such things as port fees, fuel, equipment maintenance and replacement, licensing, insurance. The amount of gross income that was profit or spent on other operating costs outside of Mexico, such as marketing, administrative overhead, staff salaries, etc., is unknown.

TABLE 2.2. Gross Earnings of Outside-Based Tourism Operators in Bahia Magdalena

Type of tour operation	Number of tourists	Typical length of tour (in days)	Typical period of stay in bay (in days)	Average price/ person (in $U.S.)	Tour operators gross earnings (in $U.S.)[a]
Cruise ships (2)	980	8	3	3,500	3,430,000
Tour boats (1)	685	8	5	1,800	1,233,000
High-price tour camps (6)	190	5	3	1,000	190,000
Low-price tour camps (4)	125	7.5	6.5	700	140,000
Tour buses[b]	300	1	0.5	250	75,000
TOTAL	2,280				5,068,000

[a] Some companies estimated their direct operating costs in Mexico to be 50–65% of their gross earnings for such things as port fees, fuel, equipment maintenance and replacement, licensing, insurance. The amount of gross income that was profit or spent on other operating costs outside of Mexico, such as marketing, administrative overhead, staff salaries, etc., is unknown.

[b] The exact number of tour busses that came to Bahia Magdalena during the 1994 gray whale season is unknown. The price/person was based on tour busses originating in the nearby cities of La Paz, Cabo San Lucas, or San Jose del Cabo. The estimated number of tourists who came to Puerto Adolfo Lopez Mateos on tour busses is based upon figures provided by the local tourism cooperative.

a growing appreciation for the skill, knowledge, and experience of local fishers in driving skiffs and observing area wildlife, resulted in foreign tour companies hiring a greater number of lagoon inhabitants in their operations. The Mexican government requires that only locals be given permits for gray whale observation from the water. Similar measures have been more recently established in Bahia Magdalena.

Although such policies are a step in the right direction, conflicts over access rights continue in both areas. For example, in Bahia Magdalena, foreign companies received government permits to operate seven whale observation skiffs with foreign drivers while local residents only received six. As a result, tourism cooperative members threatened to shut down the northern portion of Bahia Magdalena to foreign package tourism completely. This action and ensuing events received extensive coverage in the state newspaper (e.g., Firstenfeld, 1994; *Sudcaliforniano*, 1994). State government officials sponsored a meeting at the end of January 1994 in Puerto Adolfo Lopez Mateos in hopes of reducing conflict and fostering cooperation by all parties involved. As a result of the meeting, one company began to contract the services of local whale guides and a local restaurant. Other outside-based tourism companies, however, continued to operate without hiring local skiff drivers or acquiring necessary permits. Hostile exchanges between these companies and cooperative members mounted over the course of the season. But none of these companies were formally penalized through fines or otherwise.

One development that holds promise for defusing local conflicts and promoting community-based stewardship of gray whales in both Bahia Magdalena and Laguna San Ignacio has been the mobilization of nongovernmental organizations (NGOs) to protect gray whales and, more broadly, coastal marine environments around the Baja California peninsula and the Gulf of California. In the mid-1990s, the proposed development of a 52,150-hectare saltworks in and around San Ignacio Lagoon, which would be jointly owned by the Mexican government and Mitsubishi Corporation of Japan, sparked the opposition of what would become a coalition of five Mexican and United States-based NGOs: Grupo de los Cien, International Fund for Animal Welfare, Natural Resources Defense Council, Pro Esteros, and World Wildlife Fund. A key element of their strategy to defeat the development of the saltworks was to raise the awareness of local residents and the international community about the global significance of Baja's gray whale calving grounds for a migratory marine mammal that depends upon marine habitats spanning 5,000 miles and various countries (Dedina, 2000). In March 2002, the Mexican government canceled the project, with then-president Zedillo announcing that the project would change the very landscape that the biosphere reserve was meant to protect.

Since the defeat of the project, three more NGOs based in both Mexico and the United States—Pro Natura, RARE, and Wildcoast—have broadened alliances with local residents in Laguna San Ignacio, Bahia Magdalena, and others around the Baja California peninsula to expand income-generating opportunities in ecotourism while promoting more sustainable fishing activities. These groups are working to set up community-based marine protected areas; provide training for local nature guides; replace two-stroke engines for fishing and whale tour skiffs with cleaner, quieter four-stroke engines; eliminate illegal harvesting of endangered marine species; and develop community-based monitoring programs for the same (S. Dedina, personal communication, Imperial Beach, CA, 2002). While the saltworks controversy spawned a global constituency for the gray whale, it also opened a new political space for local residents in Laguna San Ignacio and other marine-dependent communities to stake a greater claim in the fate of their resources.

CONCLUSION

Ecotourism has been touted as a means to link development with conservation in the fishing communities of Laguna San Ignacio and Bahia Magdalena. Recreational whale watching may well be less destructive to marine resources than intensive harvesting of the local fishing grounds. But as Brandon and

Margoulis (1996) and Langholz (1999) point out, even if ecotourism provides a significant new source of income through environmentally friendly, non-consumptive resource use, it may not be sufficient to discourage local people from engaging in other, more destructive forms of consumptive resource use. In the two study sites, the economic benefits of gray whale tourism are not sufficient to reduce extractive pressures on inshore fisheries. Furthermore, conflicts over access to marine resources have merely intensified with the growth of ecotourism. Such conflicts undermine efforts to promote collective stewardship of marine resources among local people in both areas. This case is instructive for marine areas and other settings where ecotourism may involve the use of common-pool resources, highlighting the importance of addressing issues related to allocation of access rights and benefits from use.

The distinctive forms of gray whale tourism development that Laguna San Ignacio and Bahia Magdalena have taken also highlight the fine line that ecotourism walks as a low-impact economic activity to the environment. Because of its relative isolation, recreational whale watching has not expanded as rapidly in Laguna San Ignacio as in Bahia Magdalena. Given its high accessibility by land, air, and water, pressures to develop ecotourism and accommodate swiftly increasing numbers of tourists in the town of Puerto Adolfo Lopez Mateos are much more intense. Strains on basic services from high unregulated tourist traffic prompted the concern of some local skiff drivers, governmental regulators, and foreign tour operators in 1994. As a partial remedy to this situation, the Mexican government has since built larger restroom facilities and a permanent docking area for all whale-watching skiffs. Any attempt to devise strategies for ecologically sound income-generating activities in marine areas must consider the common-pool status of many aquatic resources as well as the contextual factors that influence local and external patterns of marine resource use (Berkes, 1985; Ostrom, Gardner, & Walker, 1994). As a multiscalar, contextual approach to understanding how markets, policies, and political processes shape nature–society relations (Blaikie & Brookfield, 1987), political ecology provides a useful framework for comparatively assessing local patterns of resource use in fishing and ecotourism along Mexico's Baja California peninsula. Politicized governance of the commons often accelerates competition among users of common-pool resources and undermines collective incentives to exercise stewardship (Brann & Foddy, 1988; Sinclair, 1990).

Conflicts over marine resources may not necessarily preclude efforts to promote conservation through ecotourism. In his study of the Maine lobster industry, Acheson (1997) showed how commercial competition and collective concern over the long-term productivity of the lobster contributed to deliberate efforts among different stakeholders to establish effective and self-enforcing regulations to protect this resource. A critical element for successful regulation of this industry has been to apportion as much power as possible to local people in governing the fishery. Unlike the case of the Maine lobster industry, however, fishing communities in Laguna San Ignacio and Bahia Magdalena are relatively new. Consequently, these communities have no locally based management practices that have evolved to regulate local use of marine resources.

Nonetheless, concerning ecotourism, area residents are aware that their economic survival depends on the survival of whales and the ecological vitality of their habitat. Furthermore, a coalition of Mexican and United States-based NGOs have opened a new space for local people to negotiate claims for access to and use of marine resources, defeating plans to develop a large-scale saltworks in Laguna San Ignacio and, more recently, engaging in a panoply of efforts to link marine conservation with both ecotourism and fishing activities. As Hanna and Smith (1993) point out, shared goals of stewardship can be a powerful means for resolving resource conflicts. The resolution of conflicts in ecotourism could serve as an important first step to resolving more entrenched conflicts in fishing and reverse present destructive trends of commercial overharvesting.

NOTES

1. As noted earlier, basic services in Laguna San Ignacio are currently nonexistent. In Puerto Adolfo Lopez Mateos, while some houses have sewage hookups, untreated effluent from sewage pipes discharges directly into the bay. Although the town has a semiregular garbage collection service, trash disposal is largely unregulated, as evident in the numerous informal garbage dumps on the outskirts of town.
2. In 1994, the price of a guided whale tour was $5/person/hour, for up to six persons/boat. This price was considerably lower than guided recreational whale watching in other parts of Bahia Magdalena as well as in Laguna San Ignacio and Laguna Ojo de Liebre, where more people were allowed onboard the whale observation skiffs for 2–4 hours for a charge of up to $30/person/hour.
3. Average reported gross earnings for 13 area residents, from package tours and other tourists arriving on their own to the lagoon, were approximately $3,100/person; net average earnings (i.e., discounting for gas and equipment maintenance) were $2,100/person.

REFERENCES

Acheson, J. M. (1988). *The lobster gangs of Maine*. New London, NH: University Press of New England.

Acheson, J. M. (1997). The politics of managing the Maine lobster industry: 1860 to the present. *Human Ecology, 25*(1), 3–27.

Aschmann, H. (1959). *The central desert of Baja California: Demography and ecology.* (Iberoamericana 42). Berkeley and Los Angeles: University of California Press.

Barkin, D. (1996). Ecotourism: A tool for sustainable development in an era of international integration? In J. A. Miller & E. Malek-Zadeh (Eds.), *The ecotourism equation: Measuring the impacts* (Yale School of Forestry and Environmental Studies Bulletin Series, No. 99, pp. 263–272). New Haven, CT: Yale University Press.

Barrett, E. M. (1974). Colonization of the Santo Domingo Valley. *Annals of the Association of American Geographers, 64,* 34–53.

Barstow, R. (1986). Non-consumptive utilization of whales. *Ambio, 15*(3), 155–163.

Berkes, F. (1985). Fishermen and the tragedy of the commons. *Environmental Conservation, 12*(3), 199–206.

Berkes, F. (1987). Common-property resource management and Cree Indian fisheries in subarctic Canada. In B. McCay & J. Acheson (Eds.), *The question of the commons* (pp. 66–91). Tucson: University of Arizona Press.

Blaikie, P., & Brookfield, H. (1987). *Land degradation and society.* London: Methuen.

Brandon, K., & Margoulis, R. (1996). The bottom line: Getting biodiversity conservation back into ecotourism. In J. A. Miller & E. Malek-Zadeh (Eds.), *The ecotourism equation: Measuring the impacts* (Yale School of Forestry and Environmental Studies Bulletin Series, No. 99, pp. 28–38). New Haven: Yale University Press.

Brann, P., & Foddy, M. (1988). Trust and the consumption of a deteriorating common resource. *Journal of Conflict Resolution, 31*(4), 6115–6130.

Brower, B. (1990). Crisis and conservation in Sagarmatha National Park. *Society and Natural Resources, 4,* 151–163.

Bryant, R. L. (1992). Political ecology: An emerging research agenda in third-world studies. *Political Geography, 11*(1), 12–36.

Carney, J. (1993). Converting the wetlands, engendering the environment: The intersection of gender with agrarian change in the Gambia. *Economic Geography, 69*(3), 329–348.

Cater, E., & Lowman, G. (Eds.). (1994). *Ecotourism: A sustainable option?* Chichester, UK: Wiley.

Ciriacy-Wantrup, S. V., & Bishop, R. C. (1975). "Common property" as a concept in natural resources policy. *Natural Resources Journal, 15,* 713–727.

Crean, K., & Symes, D. (Eds.). (1996). *Fisheries management in crisis.* Cambridge, MA: Blackwell Science.

Dedina, S., & Young, E. (1995). *Conservation and development in the gray whale lagoons of Baja California Sur, Mexico* (Final report to the U. S. Marine Mammal Commission, Contract T10155592). Washington, DC: U.S. Marine Mammal Commission.

Dedina, S. (2002). *Saving the gray whale: People, politics, and conservation in Baja California.* Tucson: University of Arizona Press.

Direccion General de Estadistica. (1972). *IX censo de poblacion de Baja California Sur, 1970.* Mexico City, Mexico: Author.

Estado de Baja California Sur. (1987). *Lineamientos de accion 1987–1993.* La Paz, Estado de Baja California Sur, Mexico: Author.

Firstenfeld, J. (1994, January 29). BCS today. *Sudcaliforniano,* p. 2-A.

Hanna, S. S., & Smith, C. L. (1993). Resolving allocation conflicts in fishery management. *Society and Natural Resources, 6,* 55–69.

Hardin, G. (1968). The tragedy of the commons. *Science, 162,* 1234–1248.

Hecht, S., & Cockburn, A. (1989). *The fate of the forest: Developers, destroyers, and defenders of the Amazon.* London: Verso.

Henderson, D. (1972). *Men and whales at Scammon's Lagoon.* Los Angeles: Dawson's Book Shop.

Hoyt, E. (1995). *The worldwide value and extent of whale watching: 1995.* Bath, UK: Whale and Dolphin Conservation Society.

Instituto Nacional de Estadistica, Geografía, y Informática. (1993). *Anuario estadístico de Baja California Sur.* Mexico City: Author.

Instituto Nacional de Estadistica, Geografía, y Informática. (1995). *Anuario estadístico de Baja California Sur.* Mexico City: Author.

Johannes, R. E. (1981). *Words of the lagoon: Fishing and marine lore in the Palau district of Micronesia.* Berkeley: University of California Press.

Jones, M. L., & Swartz, S. L. (1984). Demography and phenology of gray whales and evaluation of human activities in Laguna San Ignacio, Baja California Sur, Mexico. In M. L. Jones, S. L. Swartz, & S. Leatherwood (Eds.), *The gray whale* (pp. 159–186). Los Angeles: Dawson's Book Shop.

Langholz, J. (1999). Exploring the effects of alternative income opportunities on rainforest use: Insights from Guatemala's Maya Biosphere Reserve. *Society and Natural Resources, 12*(2), 139–150.

Lindberg, K., Enriquez, J., & Sproule, K. (1996). Ecotourism questioned: Case studies from Belize. *Annals of Tourism Research, 23*(3), 543–562.

Little, P. D., & Horowitz, M. M. (Eds.). (1987). *Lands at risk in the third world: Local-level perspectives.* Boulder, CO: Westview Press.

Maeda-Martinez, A. N. (1990). Problemas y perspectivas del cultivo de moluscos en el Pacifico mexicano. *Serie Cientifica*, U.A.B.C.S., Mexico, 1 (No. Esp. 1 AMAC), 7–12.

Marine Mammal Commission. (1994). *Annual report of the Marine Mammal Commission, calendar year 1993: Report to Congress.* Washington, DC: Marine Mammal Commission.

McCay, B. J. (1978). Systems ecology, people ecology, and the anthropology of fishing communities. *Human Ecology, 6*(4), 397–422.

McCay, B. J., & Acheson, J. M. (1987). Human ecology of the commons. In B. J. McCay & J. M. Acheson (Eds.), *The question of the commons: The culture and ecology of communal resources* (pp. 1–34). Tucson: University of Arizona Press.

McGoodwin, J. R. (1990). *Crisis in the world's fisheries: People, problems, and policies.* Stanford, CA: Stanford University Press.

Miller, J. A., & Malek-Zadeh, E. (1996). *The ecotourism equation: Measuring the impacts* (Yale School of Forestry and Environmental Studies Bulletin Series, No. 99). New Haven, CT: Yale University Press.

Mowforth, M., & Munt, I. (1998). *Tourism and sustainability: New tourism in the third world.* London and New York: Routledge.

Municipio de Comondu. (1994). *Bahia Magdalena y la ballena gris: Evaluacion del primer festival.* Ciudad Constitucion, Mexico: Author.

Nietschmann, B. Q. (1973). *Between land and water: The subsistence ecology of the Miskito Indians.* New York: Seminar Press.

Ostrom, E., Gardner, R., & Walker, J. (1994). Fishers' institutional responses to common-pool resource dilemmas. In E. Ostrom (Ed.), *Rules, games, and common-pool resources* (pp. 247–265). Ann Arbor: University of Michigan Press.

Pauly, D., Christenson, V., Guenette, S., Pitcher, T. J., Sumaila, U. R., Waters, C. J., Watson, R., & Zeller, D. (2002). Towards sustainability in world fisheries. *Nature, 418*, 689–695.

Peet, R., & Watts, M. (Eds.). (1996). *Liberation ecologies: Environment, development, and social movements.* London: Routledge.

Place, S. E. (1995). Ecotourism for sustainable development: Oxymoron or plausible strategy? *Geojournal, 35*(2), 161–173.

Romero G., I. (1990). *Efecto de las actividades humanas (lanchas turisticas y de investigacion) en el comportamiento de las ballenas grises* (Eschrichtius robustus)*: Laguna San Ignacio, Baja California Sur, Mexico.* Master's thesis, Universidad Autonoma Metropolitana, Mexico City.

Savage, M. (1993). Ecological disturbance and nature tourism. *Geographical Review, 83*(3), 290–300.

Sheridan, T. E. (1988). *Where the dove calls: The political ecology of a peasant corporate community in northwestern Mexico.* Tucson: University of Arizona Press.

Sinclair, P. R. (1990). Fisheries management and problems of social justice. *MAST, 1*, 30–47.

Stonich, S. C. (1995). The environmental quality and social justice implications of shrimp mariculture development in Honduras. *Human Ecology, 23*(2), 143–168.

Sudcaliforniano. (1994, January 21). Unicamente embarcaciones con permiso podran prestar atencion a visitantes de la ballena p. 2-C.

Villareal G., M. del Pilar. (1987). *La pesca en Baja California Sur: Mito o realidad.* Senior thesis, Universidad Autonoma de Baja California Sur, La Paz.

Western, D., & Wright, M. (Eds.). (1994). *Natural connections: Perspectives in community-based conservation.* Washington, DC: Island Press.

Wood, M. E. (1991). Global solutions: An ecotourism society. In T. Whelan (Ed.), *Nature tourism: Managing for the environment* (pp. 200–206). Washington, DC: Island Press.

Young, E. (2001) . State intervention and retreat in abuse of the commons: the case of Mexico's fisheries in Baja California Sur. *Annals of the Association of American Geographers, 91*(2), 283–306.

CHAPTER 3

Strategies for Authenticity and Space in the Maya Biosphere Reserve, Petén, Guatemala

Juanita Sundberg

Conservation organizations are attracted to environmental "hot spots" in Latin America.[1] Some protected areas attract more attention than others, either by virtue of their natural beauty or because the issues specific to the locale engender compelling narratives—Guatemala, Costa Rica, the Andes, and the Amazon basin come to mind. The sheer number of non-governmental organizations (NGOs) implementing state-supported nature reserves in Latin America begs our attention. With financial support from international donors, moral support from the scientific community, and political backing from the state, NGOs are playing an important role in managing nature and human–land relationships at multiple and intersecting political–ecological scales. Moreover, NGOs are capable of generating powerful discourses to explain environmental degradation and land use. In regions where NGOs are prevalent, local people from all socioeconomic groups necessarily interact with these discourses.

Adapted by Juanita Sundberg from the article "Strategies for Authenticity, Space, and Place in the Maya Biosphere Reserve, Petén, Guatemala," published in *Yearbook, Conference of Latin Americanist Geographers, 24* 85–86 (1998). Adapted by permission of the University of Texas Press. The adapting author is solely responsible for all change in substance, context, and emphasis.

Interested in how the relationship between NGOs and local people transforms landscapes and identities in protected areas, I conducted fieldwork in the Maya Biosphere Reserve in northern Guatemala (see Figure 3.1).[2] I was immediately struck by the appearance of conservation's new vocabulary in the many voices seeking to be heard in the reserve. This led me to hypothesize that certain individuals are appropriating conservationist discourses into their own discourses to achieve goals consistent with their own interests.[3] Drawing from participant observation and ethnographic fieldnotes, I examine in this chapter how local people (re)present themselves and their relationship to nature in new ways, thereby articulating new identities to meet changing power structures and values.

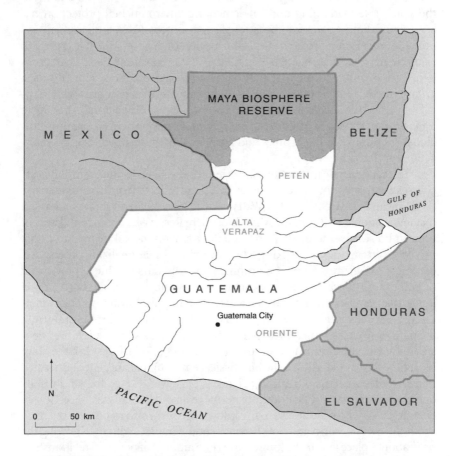

FIGURE 3.1. Guatemala and the Maya Biosphere Reserve in the Petén region.

THEORETICAL APPROACH AND METHODOLOGY

The activities of NGOs in Latin America have caught the attention of geographers (Adams, 1990; Bebbington, 1996, 1997; Bebbington & Thiele, 1993; Campbell, 2000; Price, 1994). However, few studies have examined how NGO conservation projects are being articulated in the discourses and practices of daily life at the local level (but see Zimmerer, 1996).[4] This microlevel approach is critical to understanding the politics of scale in conservation (Zimmerer, 2000). Since the 1970s, institutions operating at the global level have shaped environmental agendas the world over. Indeed, nation-states have increasingly responded to global environmental agendas by changing national legal systems to facilitate environmental protection and create protected areas; such changes directly affect local land use practices. In Latin American countries, the state's role in funding and implementing environmental protection has been significantly reduced; international environmental NGOs fill this gap and initiate conservation projects in specific locales. In this scenario, it is critical to examine how international environmental NGOs constitute local-level problems and solutions through specific discourses and practices. How do local groups interact with these discourses and practices? And how are local political ecologies transformed through interactions with global actors? Although a topic beyond the scope of this chapter, it is equally important to examine how global environmental discourses and agendas are transformed through interactions at the local level.

To examine how institutions constitute and articulate power through discourses, representations, and practices, this chapter draws from discourse analysis (Duncan & Ley, 1993; Peet & Watts, 1996a; Yapa, 1996). In this vein, I assert that international NGOs in the Maya Biosphere Reserve have generated a set of powerful discourses to explain the causes of environmental degradation that privilege certain ways of thinking while silencing or marginalizing others.[5] According to William Cronon, environmentalist discourses achieve power and moral authority because they "appeal to nature as a stable external source of non human values against which human actions can be judged without much ambiguity" (1996, p. 26). In constructing their narratives, conservationists draw from the natural sciences, which presuppose that their models of "reality" are "neutral, unbiased, objective, and value free" (Yapa, 1996, p. 711). Indeed, in the Maya Biosphere Reserve, studies of the biophysical environment are generally treated as true representations of reality.[6] Science is perceived as bias-free and wholly outside of the social, political, and economic realms of human existence. Consequently, those who have access to the knowledge seem also to possess the *truth* about the natural world and how humans should interact with it.[7] Knowing the *truth* about a place grants the knower a certain amount of power in relationships with those who are not considered to have access to that knowledge (i.e., they have not been trained in Western ways of knowing).

Moreover, conservationists' *truths* seep into policy and planning, directly impacting the lives of local people. Thus, ecology is used to back a "program of moral enlightenment" that promises to restore the "equilibrium between humans and nature" (Worster, 1990, p. 2). Indeed, NGOs in the Maya Biosphere Reserve have assumed the moral authority to speak for nature by defining which human–land relationships are compatible with the region's ecology and which are not. Relationships considered harmonious with the environment are also made to appear natural and therefore authentic, thus implying that there exists an *essential* relationship between people, place, and practices.[8] How do local people interact with these truths generated about them by others?

To examine the "conservation encounter" between NGOs and local people in the Maya Biosphere Reserve, I used ethnographic field methods.[9] I observed interactions between NGOs and locals and conducted structured and unstructured interviews with a diverse range of individuals, including NGO personnel and local people. I paid particular attention to the narratives people constructed in talking about the environment and conservation. I found that the discourse of conservation is not being used by all the reserve's inhabitants. People's class, ethnicity, gender, and past experiences interact with their articulation of identity. Those who have shown the greatest facility at coopting this vocabulary are cultural intermediaries, people who have learned to articulate multiple cultural values and practices through their relationships with ethnographers, archaeologists, individual's employed by NGOs, and other outsiders working in the Petén.

This chapter draws from my 12 months of field research focused on three case studies. The first two were located in San José, an indigenous village in the buffer zone of the Maya Biosphere Reserve. I focused on two local conservation organizations that have recently emerged in San José: one created by male leaders and another formed by women. The third case study is located in San Miguel, a migrant settlement in the heart of the reserve. San Miguel is home to the first community forestry concession, managed by the Centro Agronomico Tropical de Investigación y Enseñaza (CATIE), a Costa Rican research institution. The argument presented here reflects discourses and practices specific to the early and mid-1990s.

THE MAYA BIOSPHERE RESERVE
AND CHANGING POLITICAL ECOLOGIES

In 1990, Vinicio Cerezo, Guatemala's first civilian president elected in 15 years, signed legislation creating the Maya Biosphere Reserve to protect 1.6 million hectares of tropical lowland forest in the northern department of Petén (CONAP, 1996a; see Figure 3.1). Stretching over the southern reaches of the Yucatán's karst plateau, the reserve's vegetation includes subtropical moist and semideciduous forests, wetlands, and savannas. Important hardwoods include mahogany (Swietenia macrophylla), cedar (Cedrela mexicana), and *ceiba* (Ceiba pentandra). Species found in concentrated stands, especially around Mayan ruins, include *ramón* or breadnut (Brosimum alicastrum), *pimienta gorda* or allspice (Pimenta dioica), *copal* (Protium copal), and *chicozapote* (Manilkara zapote) (Reining & Heinzman, 1992). Nontimber forest species highly valued on the international market include chicle latex from the *chicozapote*. Also important are *xate* palms (Chamaedorea elegans and C. oblongata), ornamentals used in floral arrangements in the United States and Europe (Reining & Heinzman, 1992, p. 115). Allspice or *pimienta gorda* (Pimenta dioica) is another important export to the United States.

In 1970, an estimated 70–80% of the Petén was covered in forest. By 1986, approximately 50% of that forest cover had been felled (Schwartz, 1990, p. 12). Most of the deforestation has occurred south of the 17th parallel, as the area to the north had been protected for controlled extractive activities, such as logging, chicle extraction, and harvesting of allspice and *xate* palms.[10] The Petén's population expanded from about 25,000 in the early 1960s to 120,000 in 1978, and then to approximately 300,000–400,000 by the mid-1990s (Schwartz, 1990; Nations, 1998). About half of the migrant population are *ladinos* from the Oriente region of Guatemala and another 20% are Q'eqchi' from Alta and Baja Verapaz (SEGEPLAN, 1991, p. 39; see Figure 3.1). Deforestation has been caused primarily by cattle ranching, farming, commercial logging, and oil exploration. Over 90% of recent deforestation has occurred within 2 kilometers of roads (Sader, Sever, Smoot, & Richards, 1994).[11]

The reserve is administered by Guatemala's National Council of Protected Areas (CONAP), which, in partnership with the United States Agency for International Development (USAID), initiated a multi-million-dollar project to promote sustainable development and the "rational" management of natural resources. USAID's Maya Biosphere project is implemented through international environmental NGOs, including Conservation International (CI), the Nature Conservancy, and CARE International. The Petén has since become a mecca for NGOs—in 1996, there were over 30 (SEGEPLAN, 1996; see Figure 3.2).

The reserve's creation led to a shift in regional power structures, bringing new actors to the northern Petén with the authority to enforce a legal framework that reflects a change in how nature is constructed and valued. Prior to these changes, the principal authority in the Petén was a military-led institution, the Empresa Nacional de Fomento y Desarrollo de El Petén, or FYDEP (abolished in 1986).[12] Municipal-level politics were dominated by *alcaldes* (municipal mayors) and local military commissioners. Both the FYDEP and the municipalities supervised land use rights and regulated extractive activities like logging and chicle latex extraction (Schwartz, 1990). Aside from this, however, they did not meddle with livelihood practices.

FIGURE 3.2. NGO collaboration.

The new legal framework delineating land use in the Maya Biosphere Reserve, in contrast, directly impacts and seeks to restructure local people's relationship to the environment.[13] Many traditional activities became "infractions" or "crimes" under the new laws, including hunting, keeping certain animals as pets, and logging for domestic purposes; in certain zones, logging, gathering forest products, and slash-and-burn agriculture are forbidden (CONAP, 1996a). According to interviewees, locals tend to be affected to a much greater degree than more powerful offenders; government officials and influential figures involved in illegal logging, trafficking in animals and drugs, and looting of archaeological sites go unprosecuted.

Although CONAP has principal authority in the region, frequently its power is eclipsed by NGOs operating under the USAID's project. NGOs have no legal authority to enforce laws; rather, their authority derives from the increasing power of scientific and technological discourse to circumscribe how social groups *should* interact with nature. A host of experts has conducted studies of the Petén's biophysical environment, while the most rudimentary socioeconomic data are regarded as a sufficient source of knowledge about people's practices. For cultural and historical data, NGOs tend to rely on anthropologist Norman Schwartz's work, *Forest Society*, which seems to have acquired prescriptive status in the Petén (Schwartz, 1990).[14]

REPRESENTATIONS OF AUTHENTICITY AND LOCALNESS IN NGO DISCOURSE

The principal characters within conservationist discourse as defined by NGOs are *Peteneros* (people of the Petén) and immigrants, to the exclusion of other groups, such as ranchers, loggers, oil company employees, drug traffickers, and so on, which operate at multiple scales as they compete for access to the reserve's resources. Carmelita, Uaxactún, San Andrés, and San José are *Petenero* communities composed of farmers and forest collectors of chicle, xate, allspice, and timber. Since the first technical study conducted on the Maya Biosphere Reserve (Nations et al., 1989), NGOs have represented the forest collectors as appropriate forest dwellers for several reasons: they live within the forest and are said to be intimately familiar with it; they have a vested interest in maintaining the forest cover because it is essential to their livelihood; and it has a value to them that can be monetized. *Petenero* farmers are also considered appropriate because they tend not to engage in extensive agriculture and some have intensive agroforestry systems.

Examples of these representations are found in a local museum, the Centro dé Información de la Naturaleza, Cultura y Artesanía de Petén (CINCAP; Center for Information about Nature, Culture, and Artisans of the Petén), located in Flores. The current exhibit, installed by CI/ProPetén, represents a chicle camp, creating the impression of a simple, low-impact lifestyle. Other practices represented as "traditional" include harvesting nontimber forest products and making handicrafts. The museum also sells handicrafts made of local materials that are represented as *authentic* forest products made by *authentic* forest people. We do not learn that CI/ProPetén and a number of other NGOs funded workshops to teach people how to weave baskets and carve wooden figures (Conservation International/ProPetén, 1996).

I suggest that the museum's narratives and imagery *indigenize* or *localize* forest collectors; they become the indigenous group lacking in the reserve. I have heard NGO personnel lament the fact that the reserve does not have an indigenous group, but they have found a way around this problem by creating a narrative complete with imagery that represents forest collectors as authentic forest dwellers with authentic forest practices. While these narratives are problematic on a number of levels, it is also important to note that they transform the Petén's forest into men's space, to the exclusion of the women who inhabit and interact with it. Are women not appropriate conservation heroes?

Similarly, *Petenero* farmers are represented as members of a wise, traditional, native people. Indeed, anthropologist Scott Atran argues that the Itzaj of San José sustain a symbiotic relationship with the forest's biodiversity that has its roots in pre-Columbian practices (Atran, 1993). In outlining this close relationship, Atran's studies imply that the Itzaj are a conservationist-oriented people who have lived in harmony with the forest for at least 1,000 years. Further evidence of *Petenero* farmers' traditional virtues comes from their so-called noncommercial relationship with the forest. As Oscar, a CI/ProPetén staff member explained to me, "the *Petenero* farmer has always planted primarily for subsistence with a little extra thrown in to sell." These representations position *Petenero* farmers outside of the local economy and market.

Likewise, because some *Peteneros* maintain agroforestry systems to supplement their diet and income, NGOs attempt to encourage this practice in others. For instance, CI/ProPetén has endorsed a new conservation hero, a native of San Andrés. His extensive orchard garden, producing hundreds of pounds of fruit, is said to be exemplary of traditional practices. Hence, CI/ProPetén now employs him to promote agroforestry in other communities.[15]

So-called inappropriate forest dwellers include immigrant farmers from the Oriente and Q'eqchi' from Alta and Baja Verapaz. They are represented as inappropriate because they practice slash-and-burn cultivation; they are said to be unfamiliar with the forest; some engage in extensive agriculture for commercial purposes; and some wish to invest in cattle ranching. Although immigrants are ethnically and culturally diverse with distinct value systems, conservationist representations lump them together to create an image of destructive desperate individuals who do not think about their future—they are said to have "no commitment to the land they work" (CARE, 1996, p. 2). For instance, one conservationist explained to me that "the immigrants are not familiar with the ecosystems here in Petén, the soils are poor, and people just don't know how to manage their parcels." In the words of CARE, "The lush vegetation leads people to believe, mistakenly, that the land is extremely productive" (CARE, 1996, p. 5).

To back these highly politicized positions, NGOs rely upon scientific studies conducted by their personnel stating that the soils in the Petén are infertile and that migrants are farming on fragile lands (Nations et al., 1989; Reining & Heinzman, 1992). These studies suggest that slash-and-burn agriculture is incompatible with the tropical forest ecosystem. In articulating these positions, NGO personnel create *truths* about nature—those possessing the truth are thus able to speak for nature and determine how people should live with it.

NGO DISCOURSE AND BINARY REPRESENTATIONS

The politically charged representations described above create a set of binary oppositions between the appropriate and the inappropriate; authentic/inauthentic; wise/ignorant; and local/outsider inhabitants of the Maya Biosphere Reserve.[16] However, NGO discourses establish the authenticity of appropriate forest dwellers by what they do not say. Excluded are (at least) three intersecting factors that political ecologists emphasize in analyses of livelihood practices: individual decision making, cultural values, and historically specific socioeconomic forces (Blaikie & Brookfield, 1987; Hecht & Cockburn, 1989; Peet & Watts, 1996a; Zimmerer, 1994). In this vein, I suggest that *Petenero* relationships with the land emerge from the interaction of these factors in time and space, and are not essential cultural traits absent in others.

To back my argument, I provide a few examples from my fieldwork. The first is from Carmelita, whose inhabitants are people of mixed ethnicities and complicated histories who chose to settle in the region because the chicle industry pays well, as suggested by the local saying: *"chicleros no piden vuelta"* (*chicleros* don't ask for change). They are one of the wealthier social groups in *Petenero* society. Many come from families of *chicleros* and *cocineras* (cooks) of Mexican origin.[17] Some are infamous looters of archaeological sites deep within the forest, while others plant marijuana during the off-season (Paredes, 1997).

It is difficult to envision *chicleros* as authentic, local (in the sense of native), forest people when you consider that most migrated to the region to work in the chicle industry, which has been linked to the global trade in commodities for well over 100 years. Its collectors have fought for labor rights through a labor union established in 1948–1949, and continue to demand a higher return for their labor and a better social security system (Schwartz, 1990, p. 191). Moreover, although conservationists claim that the chicle industry does not alter the forest landscape, there is considerable evidence to the contrary. During the height of the chicle industry, over 3,000 *chicleros* and an unknown number of *cocineras* inhabited the forest, building innumerable footpaths linking chicle camps with *chiclero* routes; hundreds of hectares were cleared for corn cultivation to feed *chicleros* (Paredes, 1997; Schufeldt, 1950; Schwartz, 1990, pp. 141, 326). Finally, a certain percentage of chicle trees die each season from the extractive process (Dugelby, 1995). In short, the extractive industries have created uniquely humanized landscapes that reflect changing human–land relations.

The next example draws from my fieldwork with *Petenero* farmers, who have historically practiced slash-and-burn cultivation to grow corn, beans, and other Mesoamerican crops in plots called *milpas*. Until the 1970s, they had the advantage of large areas for crop rotation due to low population densities and usufruct land rights (Schwartz, 1990). Because the Petén's inhabitants have had difficult access to national markets, these crops were grown to meet family needs and to supply regional markets. I found that those individuals who continue to plant *milpas* use techniques that are not so different from immigrant practices, with one principal exception: immigrants tend to clear much larger areas to plant because they intend to sell to middlemen (Atran & Medin, 1997; Atran et al., 2002). Road improvements and better transportation systems now connect the Petén to Guatemala's national market.

Finally, my interviews suggest that those *Peteneros* with agroforestry systems are individuals who continue a practice that was appropriate in former times. Since times have changed, however, most have discontinued this practice for a variety of reasons. Many lost their land when the military reorganized the land tenure system in the Petén between 1966 and 1974 (Schwartz, 1990, p. 256). Some moved to towns so that their children could attend school or because they feared the violence in the mid-1980s. Moreover, some lost their trees to the mysterious rising of Lake Petén-Itza, which drowned everything they had planted around their lakeshore homes. Others have dedicated their time to nonagricultural occupations. Still others have found that local demand is filled by produce from other parts of the country. Those who continue to expand their orchard gardens are individuals that enjoy this occupation— *"Le gusta sembrar"* (She or he likes to plant"), as they say.

Many immigrants stated that they chose not to adopt agroforestry practices on an intensive basis because, according to their calculations, the benefits do not outweigh the costs in terms of time, transportation, and sales to middlemen. Nonetheless, I have heard NGO personnel describe these individuals as lazy, ignorant, and incapable of thinking ahead—attitudes said to be inherent to their culture. One extenuation agent in San Miguel said that immigrants who did not adopt his project's agroforestry practices "do not have a culture of planting trees." In this case, it was more an issue of technique, aesthetics, and market considerations than cultural behavior. The NGO was promoting monoculture citrus orchards (planted in straight lines) designed for market production (see Figure 3.3). Migrant farmers resisted these efforts and continued to work in the smaller scale, mixed-use orchard gardens around their residences and in their *milpas*—which the extension agent did not recognize as legitimate forms of agroforestry. Aware of the market limitations of expanding production, individuals chose to plant for local use, including barter and gift giving. Finally, in contrast to monoculture orchards, migrant orchard gardens promote and sustain biodiversity, preserve genetic stock of native species, and provide habitat and food sources to birds, insects, and foraging animals (Atran, 1993; Ferguson et al., 2003).[18] If and when they are abandoned, mixed-use orchard gardens allow for and may even promote forest regeneration.

In sum, the culture and practices of *Petenero* forest collectors and farmers emerged in specific social, historical, and spatial contexts. As times have changed, so have people's practices. Conservationists, however, have looked to

FIGURE 3.3. Tree culture.

the past to prescribe the future through the discursive construction of a *Petenero* traditional cultural ecology, which is then taught to immigrant communities as appropriate practice. In the following section, I explore how people living in the reserve interact with these representations.

CONSERVATION AND THE POLITICS OF LOCAL IDENTITIES

Conservation is articulated at multiple sociopolitical scales; attention to the micropolitics of conservation at the local level reveals the complex intersections between donor agendas, NGO projects, and local concerns. In the Maya Biosphere Reserve, conservationist discourses often embody *Petenero*–immigrant tensions, in that the *Peteneros* feel that their land has been invaded and destroyed by outsiders. Consequently, people who consider themselves *Peteneros* are more likely to express an affinity for NGOs. The words of Don Luis of San José exemplify a commonly heard assertion: "*We* knew how to take care of the forest," he said. In interviews, many told me that *Peteneros* know how to care for the forest, whereas the immigrants have come to destroy it. There is a great deal of resentment and even hatred between groups; immigrants are commonly vilified. Educated *Peteneros* whose families once collected forest products and farmed small areas now proclaim themselves as the legitimate caretakers of the forest, often implying that *they* should have increased rights over its management. Most no longer depend upon these activities for their livelihood; however, their comments often reflect resentment of the period when many lost land due to immigration and subsequent privatization of land tenure by the FYDEP in the 1960s and 1970s.

Many educated *Peteneros* have sought to reaffirm their rights over the Petén's forest by working in conservation organizations or government institutions, which gives them a say in policy and project design. My interviews suggest that *Peteneros* have selected those aspects of international conservation discourse that meld neatly with their own attitudes. I found their resentment against migrants to be made manifest in their discourse, their behavior, and even in their project orientation.

One plan designed to give locals more control over their forests is the community forestry concession, which is also intended to promote sustainable development. In Carmelita and Uaxactún, forest collectors understand that conservationists value their livelihood practices and knowledge of the forest. I found that particular leaders frequently appropriate this discourse to their own ends and reinvent themselves as harmonious forest dwellers and conservation heroes. This strategy supports their long-term desire for more control over areas of traditional use to keep out immigrants and to ensure that they, and not illegal loggers or the logging industry, benefit from timber and nontimber extraction.

Whenever he has an audience, Don Carlos, an outspoken and controversial leader from Carmelita, will explain that the forest collectors know how to take care of the forest because they belong to a "forest culture."[19] He therefore demands community rights to control extraction, immigration, and especially industrial timber harvesting. To a group of advisers and extension agents, he said "You people do your work from the office. We are the ones out in the field. We know the situation the wildlife is in. We want to protect the forest because we live in it." In short, *Peteneros* such as Don Carlos are able and willing to play with the ways in which conservation articulates at various scales: understanding how conservation agendas intersect with local politics gives them a new arena in which to assert power through claims to locality.

In the case of San Miguel, a migrant settlement, residents are the beneficiaries of a community forestry concession, but not because of their knowledge of the forest or even their traditional livelihood practices. Rather, after the neighboring town refused to consider a forestry concession in 1990, CATIE, a Costa Rican institution, began to approach community members in San Miguel about their interest in a concession. Up to this point, the 30 families in the San Miguel area faced an uncertain future within the newly declared reserve, and many feared they would be forced out. All are immigrants to the Petén, although the majority have resided in San Miguel for 10–15 years. Despite their reservations and all-around suspicion that CATIE was trying to trick them into giving up their land, a group of male leaders agreed to form a committee and establish the forestry concession.

My interviews in San Miguel reveal that people support the concession not because they are interested in forestry, or even because they think that it will be financially rewarding, but because it provides them with legal assurance that they will not be thrown off their land and out of the reserve. As one man commented, "The land is [now] ours and we are paying taxes to harvest. We are renting the land, which is 5,000 hectares for 40 years, and we are going to harvest 150 hectares per year." Moreover, it enables them to control the extraction of resources within the concession's boundaries, thereby ensuring that they will benefit, not others. Finally, the concession allows them to keep other migrants from moving into the area. This is indicated in Chema's assertion that: "we know that we are renting this land and that they can't remove us, nor can others come in."

Although the concessions were designed to establish community management of space and resources, the CATIE staff is reluctant to relinquish control to local people on the presumption that locals do not have the knowledge or capacity to take on the mental tasks. The staff's training in agronomy or forestry gives them a certain amount of authority—if knowledge is defined as the result of scientific training, then its absence is assumed among the uneducated. Locals also see themselves as lacking this mysterious quality and the knowledge that they *do* have is disregarded. Consequently, local men tend to represent themselves as incapable of managing the concession. This way of legitimating knowledge—as that which pertains to the educated–feeds into Guatemala's entrenched class system.

During the logging season in San Miguel, I found that the CATIE staff perpetuated this class-based distinction, which effectively discourages a local sense of ownership or involvement in the project. I observed a distinct division of labor: the technical staff did the mental work and the community members did the physical labor. When felling and loading timber, for example, the staff delegated tasks to community members, who receive a daily wage that comes out of the committee's earnings. *San Migueleños* regard this as simply another form of day labor. Ironically, the project director lamented to me that the committee members "just don't see the project as their own."

My findings in San Miguel exemplify what a Guatemalan development consultant calls the "culture of simulation," which he has frequently observed in development projects. As he put it, the technical staff accomplishes the goals set out in the management plan, thereby "simulating" that they are helping the locals. The locals participate in the projects and do as they are told, thereby "simulating" their process of development. In the meantime, "things remain the same." In this case, I would argue that the simulation tactics are part of *San Migueleños'* strategy to get what they want from the project. It will be some time before CONAP and participating NGOs are able to determine whether or not the community forestry concessions fit their model of sustainable development. In the meantime, the articulation between global conservation agendas and local interests enable people in San Miguel to obtain legal rights to place, space, and livelihood.

NGO DISCOURSES, NGO LANDSCAPES

In this chapter, I argue that NGOs, as international institutions seeking to implement global environmental agendas, have the power to construct *truths* about human–land relationships. In assuming the moral authority to speak for nature, NGOs discourses frame the way environmental degradation is conceptualized at the local level. Certain perspectives are privileged over others and then worked into national political and legal structures to delineate how local people should relate to the environment. Yet NGO discourses are not objective, bias-free mirrors of reality; they are embroidered within the very fabric of power structures at local, national, and global scales. Consequently, the natural and social sciences are used to prescribe culturally constructed preferences that risk supporting/creating social structures that, in turn, perpetuate ethnic-, class-, and gender-based inequalities.

In the Maya Biosphere Reserve, local people at every socioeconomic level necessarily interact with NGO discourses. In this chapter, I suggest that certain groups have begun to draw from the discourses of conservation to reposition themselves in relation to NGOs. For instance, long-standing social divisions, such as *Petenero*-immigrant struggles, are reworked within conservationist discourses. As shown, *Peteneros* seek to establish their legitimacy as caretakers of the Petén's forest through discourses of authenticity and locality that draw from and feed into North American idealizations of the *bon sauvage*. Lacking this authenticity, immigrants in San Miguel participate in the community forestry concession to achieve their goal of land tenure security. As they struggle to negotiate changing power structures at multiple scales, both groups reinvent localized identities to assert rights to space, place, and livelihood.

Appropriating powerful discourses and reinventing themselves as forest dwellers serves the interests of particular groups in the Petén. The prickly question that continues to arise relates to those groups increasingly marginalized by conservation projects. In a country like Guatemala, where the state has little funding for social services, what happens to those people whose livelihood practices are deemed inappropriate by conservation organizations?

RESEARCH AND POLICY IMPLICATIONS

Political ecologists must be attentive to the ways in which socially constructed notions of cultural traits and practices are tied to power relations at multiple scales. Studies by NGOs, for instance, must be situated within a politicized field of knowledge. Political ecologists will be well positioned to analyze how global conservation agendas articulate with local political ecologies if we begin with the assumption that human–land relationships emerge within particular socioeconomic contexts in time and space. Such research, when used by policymakers, may help to foresee how unequal power relations are woven into decision-making processes and policies, with uneven consequences for differing social groups.

NOTES

1. See, for instance, Conservation International's web page at *http://www.conservation.org*, which features "hot spots."
2. Funding for fieldwork in the Maya Biosphere Reserve from February 1996 to March 1997 was provided by an Institute of International Education Fulbright Fellowship and supported by a Dissertation Travel Grant from the AAG Latin American Specialty Group and Travel Grants Committee. This chapter represents a small nugget extracted from the research I conducted. I wish to thank Gregory Knapp, Karl Offen, Thomas Bassett, and Karl Zimmerer for helpful suggestions. This chapter would not have been possible without the participation of the women and men residing within the Maya Biosphere Reserve, who offered to share their perspectives with me. I am responsible for the interpretation and argument, including any errors found herein.
3. By "conservation's new vocabulary," I am referring to an international, standardized discourse about ecosystems and nature protection (McAfee, 1999). This discourse is most visible in documents produced by institutions such as the United Nations, the World Bank, and the United States Agency for International Development. Sachs (1992) and Escobar (1996) argue that this discourse expresses an economistic approach to nature that is foreign to many societies around the world. Examples of this vocabulary include terms like "ecosystem," "natural resources," "nature protection," and even the word "conservation."
4. Even fewer studies examine how NGOs are effecting changes in the biophysical landscape at various scales.
5. My argument draws from Michel Foucault's (1978) analyses of the relations between institutions, discourse, and material practices. See Peet and Watts (1996a), for an extended discussion of how this approach can be useful in political ecology.
6. The studies I am referring to are conducted by NGO consultants or staff to augment knowledge of the region. However, several important studies conducted prior to the creation of the reserve have had an influence on the discourse that fol-

lowed (Nations et al., 1988, Nations et al., 1989, and Reining & Heinzman, 1992, are widely cited; other studies by the extractive industries and research and development institutions are cited to a lesser degree).

7. See Sundberg (1998) for a fuller discussion of conservationist truths about the Maya Biosphere Reserve.

8. The process of "essentializing" involves reducing an individual or a group's memorable or recognizable traits to fixed (or permanent) characteristics, which are then naturalized through discourse (Fuss, 1989).

9. There are many different kinds of NGOs working in the Maya Biosphere Reserve. My research touches on three private, professionally staffed, North American organizations that have contracts with the USAID to implement the Maya Biosphere Project: The Nature Conservancy, Conservation International (CI), and CARE. CI has created a local NGO, CI/ProPetén, which was dependent on CI Washington for funding and policy design until 2002. A fourth NGO is CATIE, Centro Agronomico Tropical de Investigación y Enseñaza or Center for Education and Investigation in Tropical Agronomy, a private, nonprofit, educational institution based in Turrialba, Costa Rica. When I am referring to NGO discourses in general, I am mostly alluding to these organizations, along with a few others, like Centro Maya, the Peregrine Fund, and the World Conservation Society, as they have more financial, political, and moral support than other institutions.

10. The Empresa Nacional de Fomento y Desarrollo de El Petén (FYDEP, or the National Agency for Promotion and Development of the Petén), informally designated the region north of parallel 17′ 10° as a forest reserve for selective logging and extraction of nontimber products; the institution attempted to prevent agricultural settlements therein (Schwartz, 1990, p. 253; Soza Manzanero, 1996, p. 22). The Maya Biosphere Reserve encompasses this forest reserve. The FYDEP was a military-led, government institution given "extensive and in practice exclusive authority" in the Petén, ostensibly to promote economic development in the region (Schwartz, 1990, p. 253).

11. According to Sader et al. (1994, p. 325), the "greatest amount of forest clearing between 1986 and 1990 occurred along the road from La Libertad to El Naranjo in the southwest." This is an old and long-used road from central Petén to the western border with Mexico. Another major region of deforestation is "adjacent to the western border with Mexico" (Sader et al., 1994, p. 322).

12. See note 10.

13. The Maya Biosphere Reserve Law establishes zones with laws regulating land use in the different zones, following UNESCO's vision of how biosphere reserves should function to unify conservation with sustainable development (CONAP, 1996a, 1996b; UNESCO, 1984).

14. *Forest Society: A Social History of Petén, Guatemala*, is the only English language text of its kind; it has been widely read and cited by people working in the Petén. The title of the book has taken on a life of its own, independent of the author and his intentions. I contend that it has acquired "prescriptive status" in the sense that, in current discourse, society in the Petén is represented as authentic only if it is a "forest society." I wish to make it clear that when I reference the term "forest society" I am referring to its place in discourse, independent of the author who penned

the term; my usage is not meant as a reference to Schwartz himself or his scholarship.

15. See Grandia, Reining, and Soza Manzanero (1998, p. 369) for a description of this individual.

16. Here my argument draws from theories that view meaning as constructed by the creation and perpetuation of difference through binary oppositions. From this perspective, identity is continuously constructed through interactions with an Other and categorizations of the Other always draw from and construct a picture of the self. I am drawing from Foucault (1978) and Said (1979), who themselves draw from ideas developed by F. de Saussure, S. Freud, J. Derrida, C. Lévi-Strauss, J. Kristeva, and J. Lacan, to name just a few.

17. The word cocinera refers to a woman hired to cook for the *chiclero* camps. This is a difficult but well-paying job.

18. Orchard gardens commonly include *sapotes*, hogplums, *anonas*, avocados, exotic fruit trees, and a host of useful timber species (Atran, 1993, p. 640). *Milpas* may include maize, squash, chile peppers, beans, tomatoes, sweet potatoes, root crops, plantains, cotton, manioc, tobacco, as well as timber species (Atran, 1993).

19. Carlos Catalan was assassinated in June 1997 while transporting Carmelita's community forestry concession's first timber harvest.

REFERENCES

Adams, W. M. (1990). *Green development, environment and sustainability in the third world.* London: Routledge.

Atran, S. (1993). Itza Maya tropical agro-forestry. *Current Anthropology, 34*(5), 633–699.

Atran, S., & Medin, D. (1997). Knowledge and action: Cultural models of nature and resource management in Mesoamerica. In M. Bazerman (Ed.), *Environment, ethics, and behavior: The psychology of environmental valuation and degradation* (pp. 171–208). San Francisco: New Lexington Press.

Atran, S., Medin, D., Ross, N., Lynch, E., Vapnarsky, V., Ucan Ek', E., Coley, J., Timura, C., & Baran, M. (2002). Folkecology, cultural epidemiology, and the spirit of the commons: A garden experiment in the Maya lowlands, 1991–2001. *Cultural Anthropology, 43*(3), 421–450.

Bebbington, A. (1996). Movements, modernizations, and markets. In R. Peet & M. Watts (Eds.), *Liberation ecologies, environment, development, social movements* (pp. 86–109). London and New York: Routledge.

Bebbington, A. (1997). New states, new NGOs?: Crises and transitions among rural development NGOs in the Andean region. *World Development, 25*(11), 1755–1765.

Bebbington, A., & Thiele, G. (1993). *Non-governmental organizations and the state in Latin America: Rethinking roles in sustainable agricultural development.* London: Routledge.

Blaikie, P., & Brookfield, H. (1987). *Land degradation and society.* London: Methuen.

Campbell, L. (2000). Human need in rural developing areas: Perceptions of wildlife conservation experts. *Canadian Geographer, 44*(2), 167–181.

CARE. (1996). *The Maya Biosphere Reserve project.* Sta. Elena, Petén, Guatemala: Author.

CONAP. (1996a). *Ley de Areas Protegidas y Su Reglamento.* Guatemala City, Guatemala: Author.

CONAP. (1996b). *Plan Maestro: Reserva de la Biósfera Maya.* Turrialba, Costa Rica: Author.

Conservation International/ProPetén. (1996). *Sus Actividades en la Reserva de la Biósfera Maya.* Flores, Petén, Guatemala: Author.

Cronon, W. (1996). Introduction. In W. Cronon (Ed.), *Uncommon ground. Rethinking the human place in nature* (pp. 23–56). New York: Norton.

Dugelby, B. (1995). *Chicle latex extraction in the Maya Biosphere Reserve: Behavioral, institutional and ecological factors affecting sustainability.* Unpublished doctoral dissertation, Duke University, Durham, NC.

Duncan, J., & Ley, D. (Eds.). (1993). *Place/culture/representation.* London: Routledge.

Escobar, A. (1996). Constructing nature. In R. Peet & M. Watts (Eds.), *Liberation ecologies, environment, development, social movements* (pp. 46–68). London and New York: Routledge.

Foucault, M. (1978). *The history of sexuality.* New York: Pantheon Books.

Ferguson, B., Vandermeer, J., Morales, H., & Griffith, D. (2003). Post-agricultural succession in El Petén, Guatemala. *Conservation Biology, 17*(3), 818–828.

Fuss, D. (1989). *Essentially speaking: Feminism, nature, and difference.* New York: Routledge.

Grandia, L., Reining, C., & Soza Manzanero, C. (1998). Illuminating the Petén's Throne of Gold: The ProPetén experiment in conservation-based development. In R. Primack (Ed.), *Timber, tourists, and temples: Conservation and development in the Maya Forest of Belize, Guatemala, and Mexico* (pp. 365–388). Washington, DC: Island Press.

Hecht, S., & Cockburn, A. (1989). *The fate of the forest: Developers, destroyers and defenders of the Amazon.* London: Verso.

McAfee, K. (1999). Selling nature to save it?: Biodiversity and green developmentalism. *Environment and Planning D: Society and Space, 17*(2), 133–154.

Nations, J. (1998). The uncertain future of Guatemala's Maya Biosphere Reserve. In J. Nations (Ed.), *Thirteen ways of looking at a tropical forest* (pp. 10–13). Washington, DC: Conservation International.

Nations, J., Billy, S., Ponciano, I., Houseal, B., Castillo, J. J., Godoy, J. C., & Castro, F. (1989). *La Reserva la Biósfera Maya, Petén: Estudio técnico.* Guatemala City, Guatemala.

Nations, J., Houseal, B., Ponciano, I., Billy, B., Godoy, J. C., Castro, F., Miller, G., Rose, D., Rey Rosa, M., & Azurdia, C. (1988). *Biodiversity in Guatemala.* Washington, DC: World Resources Institute, Center for International Development and the Environment.

Paredes, S. (1997). *Surviving in the rainforest: The realities of looting in the rural villages of El Petén, Guatemala* (Monograph). Crystal River, FL: Foundation for the Advancement of Mesoamerican Studies.

Peet, R., & Watts, M. (1996a). Liberation ecology. In R. Peet & M. Watts (Eds.), *Liberation ecologies, environment, development, social movements* (pp. 1–45). London and New York: Routledge.

Peet, R., & Watts, M. (Eds.). (1996b). *Liberation ecologies, environment, development, social movements.* London and New York: Routledge.

Price, M. (1994). Ecopolitics and environmental nongovernmental organizations in Latin America. *Geographical Review, 84*(1), 42–59.

Reining, C., & Heinzman, R. (1992). *Non timber forest products of the Maya Biosphere Reserve, Petén, Guatemala.* Washington, DC: Conservation International Foundation.

Sachs, W. (1992). *The development dictionary: A guide to knowledge as power.* London: Zed Books.

Sader, S. A., Sever, T., Smoot, J. C., & Richards, M. (1994). Forest change estimates for the northern Petén region of Guatemala, 1986–1990. *Human Ecology, 22*(3), 317–332.

Said, E. (1979). *Orientalism.* New York: Vintage Books.

Schufeldt, P. W. (1950). Reminiscences of a *chiclero.* In A. Anderson (Ed.), *Morleyanna: A collection of writings in memorium of Sylvannus Griswold Morley, 1883–1948* (pp. 224–229). Santa Fe: School of American Research and Museum of New Mexico.

Schwartz, N. (1987). Colonization of northern Guatemala: The Petén. *Journal of Anthropological Research, 43,* 163–83.

Schwartz, N. (1990). *Forest society: A social history of Petén, Guatemala.* Philadelphia: University of Pennsylvania Press.

SEGEPLAN. (1991). *Informe de Orientación. Analisis demográfico y sociológico. Plan de Desarrollo Integrado de Petén.* Santa Elena, Petén, Guatemala: Author.

SEGEPLAN. (1996). *Directory of government and non-governmental institutions in the Petén.* Santa Elena, Petén, Guatemala: Author.

Soza Manzanero, C. (1996). *Factores que inciden en la Conciencia Ecológica de los Habitantes de la Reserva de la Biósfera Maya en el Departamento de el Petén.* Guatemala City, Guatemala: Departamento de Pedagogia, Facultad de Humanidades, Universidad de San Carlos de Guatemala.

Sundberg, J. (1998). NGO landscapes: Conservation and communities in the Maya Biosphere Reserve, Petén, Guatemala. *Geographical Review, 88*(3), 388–412.

UNESCO. (1984). Action plan for biosphere reserves. *Nature and Resources, 2*(4), 11–22.

United States Agency for International Development. (1989). *Maya Biosphere Project* (Project Paper). Washington, DC: Author.

Worster, D. (1977/1995). *Nature's economy: A history of ecological ideas.* Cambridge, UK: Cambridge University Press.

Worster, D. (1990). The ecology of order and chaos. *Environmental History Review, 14*(1–2), 1–18.

Yapa, L. (1996). What causes poverty?: A postmodern view. *Annals of the Association of American Geographers, 86*(4), 707–728.

Zimmerer, K. (1994). Wetland production and smallholder persistence: Agricultural change in a highland Peruvian region. In K. Foote (Ed.), *Re-reading cultural geography* (pp. 260–280). Austin: University of Texas Press.

Zimmerer, K. (1996). Discourses on soil loss in Bolivia. In R. Peet & M. Watts (Eds.), *Liberation ecologies, environment, development, social movements* (pp. 110–124). London and New York: Routledge.

Zimmerer, K. (2000). The reworking of conservation geographies: Nonequilibrium landscapes and nature–society hybrids. *Annals of the Association of American Geographers, 90*(2), 356–369.

PART II

URBAN AND
INDUSTRIAL ENVIRONMENTS

CHAPTER 4

Toward a Political Ecology of Urban Environmental Risk

The Case of Guyana

Mark Pelling

At the turn of the second millennium the balance of the world's population shifted from being rural to being urban, with the majority of the world's urban populations living in the so-called developing countries of Africa, Asia, and Latin America. By 2005, for every one urbanite living in a developed country, two will reside in developing countries, with this ratio set to rise to 1:3 around 2025 (United Nations, 1989). As urban populations have grown and the consumption and production of goods has become increasingly concentrated in or driven by urban industrial centers, we have become more aware of the role of cities as engines for transforming the environment and as places of great vulnerability to environmental risk. Political ecology has a role to play in exploring both of these themes, but it is the latter that is examined in this chapter.

The political ecology of urban environmental risk builds on a vibrant tradition of political analysis in urban studies. In particular, work on sustainable urbanization (see the journal *Environment and Urbanization*) and environmental justice (e.g., Agyeman, 2000) shares with rural political ecology a critically engaged yet constructive worldview and an interest in political economic

Adapted by Mark Pelling from the article "The Political Ecology of Flood Hazard in Urban Guyana," published in *Geoforum*, *30*(3), 249–261 (1999). Adapted by permission of Elsevier Science. The adapting author is solely responsible for all changes in substance, context, and emphasis.

explanations for the social distribution of the environmental risks and benefits of development. Political ecology explanations of risk can usefully draw on the human ecology school of hazards analysis (Blaikie, Cannon, Davis, & Wisner, 1994; Hewitt, 1983), which conceptualizes environmental risk as the coincidence of physical hazard and human vulnerability (see also Swyngedouw, Chapter 5, this volume). Notwithstanding the increased, albeit uneven, adoption of urban sustainability measures, such as the Agenda 21 guidelines, political ecological vulnerability remains "intimately connected with the continuing process of underdevelopment recorded throughout the world" (O'Keefe, Westgate, & Wisner, 1976, p. 560).

In this chapter, discussion of a political ecology framework for the study of urban environmental risk is followed by an illustrative account of urban flooding in Guyana. More than 90% of Guyana's urban population and 75% of its GNP-producing activities are at risk from coastal and runoff flooding (Pelling, 1996). The historical root causes of risk can be traced to Guyana's colonial experience and to postcolonial modernization projects that transformed the coastal environment by clearing and replacing coastal mangrove stands with a landscape of sea walls, irrigation canals, plantations, and human settlements. This created "second nature" (Smith, 1984) requires high levels of human inputs (both labor and financial) to maintain. The vulnerability of the settled coast to flooding and the future impacts of climate change are explained by the failure of the contemporary political economy of the coast to produce itself or to access from external sources the inputs required for its maintenance. The influence of political structures and cultural norms on the social and spatial distribution of risk is explored in the case study.

THE POLITICAL ECOLOGY OF RISK IN URBAN ENVIRONMENTS

Political and socioenvironmental analysis are placed at the center of analysis in bringing a political ecology framework to the study of urban environmental risk. Political power operates on two levels: in the construction of discourses and as the means for shaping the material and social world. Competition between political actors for control of discourse is often a precondition for control of the mechanisms of resource distribution (Peet & Watts, 1996). Competition for control of political discourse on the urban environment is well exemplified by policy debate over solid waste as a livelihood resource or a health hazard, and over squatter housing as a solution to low-income housing demand or a social ill. Such differences in perceptions of the urban environment are often experienced as a conflict between popular and technical–scientific knowledge systems and their related political constituencies. Persons and institutions that control the content of political discourse in the city determine the legitimacy and prioritizing of development policy options. Enacting these privileged policies requires power in the form of human skills and financial resources. Of course, in the city there is never only one discourse or a single set of privileged policy choices. The tolerance or silencing by the city's political–business–administrative elite of alternative and competing projects is an indicator of the openness, responsiveness, and flexibility of the urban polity, and potentially of the city's sensitivity to risk and its capacity for adaptation (Wildavsky, 1988).

Urban societies are more vulnerable than their rural counterparts to environmental risk, as has been argued with reference to the cities of Africa, Asia, and Latin America (Moser, Gauhurst, & Gonan, 1994). This difference in risk vulnerability is due to the commodified nature of the urban economy (integration into a cash economy), a greater variety of environmental dangers (including toxic contaminants, industrial pollution, natural disasters, and epidemic disease), and social fragmentation (city dwellers' loss of supportive social networks, greater social problems). This view contradicts established wisdom, which considers urban populations to be more secure from risk because of the greater availability of services, infrastructure, and economic resources in urban areas (Drèze & Sen, 1989). Clearly, the context of risk is critical to the production of relative vulnerabilities. Comparing urban areas, McGranahan, Jacobi, Songsore, Surjadi, and Kjellén (2001) delineate an environmental transition, with less wealthy communities and cities at risk from poor sanitation, medium-ranking settlements exporting waste but remaining vulnerable to air pollution, and high-wealth cities contributing to global environmental change. In the urban milieu the actions of individuals and groups in one settlement can rapidly and profoundly influence the security of those living in other places and in later times.

Vulnerability to environmental risk for individuals and social groups has three components: exposure to risk, resistance to the impacts of a hazard, and capacity to adapt to reduce exposure to future risks (Pelling, 2001). Adaptive capacity is key for avoiding future risks and is rooted in an agent's ability to compete for access to rights, resources, and assets (Sen, 1981; Blaikie et al., 1994), and takes on their fluctuating status. There is an overlap between experiences of poverty and vulnerability, but they are not equivalent terms. Poor households constantly have to play off poverty and vulnerability (Swift, 1989), invariably "choosing" to accept greater vulnerability (a lack of long-term investment in housing quality, education, or health care) in their daily battle for survival amid poverty (Chambers, 1995). The poor are often vulnerable, but the vulnerable are not always poor. For example, in the 1999 Marmara earthquake in Turkey the greatest losses were suffered by middle-income residents of inappropriately constructed apartment blocks (Özerdem, 2003). The remainder of this section highlights debates over the pathways through which human vulnerability to urban environmental risk might be produced.

Echoing explanations of rural environmental degradation, rapid population growth is frequently presented as a cause of rising urban environmental risk. As in accounts of rural development, such an approach is in danger of relying on an uncritical neo-Malthusian framework of analysis (Bryant, 2001). As city populations grow, competition for living space within the city intensifies and the volumes of resources consumed and waste produced increases. But wealth, culture, administrative capacity, and political orientation are important factors too. In cities with high levels of income inequality and in the absence of support for large-scale social housing schemes, growing numbers of people are excluded from the formal land and housing markets, forcing many to live in unplanned and hazardous places. In rich and poor cities alike it is common for urban residents to suffer this kind of exclusion and resulting environmental injustice. But the proportions of urban populations forced to choose to live and work in places of environmental risk is highest in Africa, Asia, and Latin America. In Bogota, 60% of the population live on steep slopes subject to landslides, and in Calcutta 66% of the population live in squatter settlements at risk from flooding and cyclones (Blaikie et al., 1994).

Hewitt (1997) argues that the density of urban living contributes to risk. Cities make up only 1% of the land area of the earth, but estimates suggest that they concentrate more than half of the world's population and the major portion of its physical capital (buildings, infrastructure). Hewitt (1997) proposes that concentration leads to vulnerability through a number of reinforcing pathways. First, risk arises from the increasing scales and concentration of energy or energy transportation routes and their proximity to residential areas. Second, living in crowded conditions increases risk through the likelihood of disease transmission and congestion, which constrains disaster relief. Third, cities are complex places with many interactions between different elements— some less predictable than others. Fourth, during periods of political misrule urban populations often become easy targets for exploitation or repression. Lagos, Nigeria, has a history of military authoritarianism and violence. The 2002 accidental explosion of an army weapons dump in Lagos demonstrates well the vulnerability of urban populations to disaster. Here a fire in an ammunition dump led to 700 deaths in neighboring residential areas. The army had time to evacuate the site but did not warn local residents. Most of the dead were not killed in the blast but drowned in a canal while trying to escape (Pelling, 2003).

Resource scarcity for urban administrations and a burgeoning population that outstrips supply has been exacerbated in many cities by a failure to organize strategic urban planning that is accountable to and informed by open forms of urban governance. Despite being one of the largest cities in the world, Mexico City has endured many years with no effective urban planning. Garza (1999) links this problem to the structural adjustment experience, reflecting the neoliberal vision of urban development propagated by Western financial institutions that fits easily with a presumption that the real estate market and not planning is the proper tool for determining land use and urban morphology. A similar situation was found in Monterrey, where despite the legal requirement for consultation on development plans, Garza (1999) argues that nongovernmental organizations (NGOs) and other popular organizations were excluded and the planning committee was captured by real estate interests. Cities need to attract global capital investment to support social and environmental planning. However, the footloose nature of global capital means that companies can shift risk down the supply chain or onto labor through job insecurity, fluctuating incomes, and a lack of unionization (Mabey & McNally, 1999). Following the Kobe earthquake in Japan, the branch plants of multinational car manufacturers were able to re-source the accessing of components from damaged companies. Thus the multinationals contained their production losses but the smaller component manufacturers were hit twice: once by earthquake damage and then again by lost revenue. There is a pressure for cities to capitalize on their competitive advantages to attract foreign direct investment by turning a blind eye to poor employment practices. There is also pressure for established companies, large and small, to cut costs to maintain their market share. This is too easily translated into an erosion of worker and industrial safety, environmental management, and social development. The toxic gas leak in Union Carbide (India) Ltd's plant in Bhopal in 1984 can be explained by just such a cocktail of causes. In total some 3,500 people were killed and 300,000 were injured (Pelling, 2003).

The technology exists to map environmental risks, build safe houses, and plan for secure cities. So why does this not happen? In a review of hazards management in the United States, White, Kates, and Burton (2002) suggest that urban risk arises not out of a lack of technical expertise but rather because of intervening sociopolitical structures that undermine or misshape attempts at mitigation policy. Thus flood control projects actually serve to encourage floodplain development and land use restrictions are watered down or overruled by political means. In hazardous areas conflicting interests and a lack of political will to resolve them seems to be at the base of many failures to apply knowledge effectively. If this accurately describes the sociopolitical barriers to disaster planning in the United States, it must only scratch at the surface of the difficulties encountered by urban planners with responsibility for disaster mitigation in the cities of Africa, Asia, and Latin America that have to operate within much tighter financial and human skill constraints. In concluding his review of disaster and risk management in 10 megacities from around the world Mitchell (1999, p. 480) makes the following observations:

> The neglected approaches involve non-expert systems, informal procedures, non-structural technologies, private sector institutions, and actions taken by individuals, families, neighbourhood groups, firms, and similar entities. . . . There

is a lack of initiatives that jointly address different kinds of hazard, a slowness to integrate hazards management with other problem-solving urban programmes, and a failure to investigate other roles that hazard plays in the lives of urban residents.

At the heart of these observations lies a recognition of the need for more awareness of the action of informal, or nonstate, actors. This should *not* be seen as a call for shifting attention—and financial/technical support—from the public sector, but rather as an argument for broadening the list of actors and institutions that are included in planning for urban disaster. In the end this list might more closely reflect the types of institutions that vulnerable individuals and groups interact with in attempts to deal with their own exposure to risk (Pelling, 2003).

APPLYING A POLITICAL ECOLOGY FRAMEWORK: URBAN RISK IN GUYANA

Historicizing Power and Vulnerability

The Guyanese polity has been divided along ethnic lines since shortly after independence in 1966, and this division lies at the heart of contemporary environmental injustice. The two main parties are the People's Progressive Party (PPP), with rural and Indo-Guyanese support, and the People's National Congress (PNC), with a more urban and largely Afro-Guyanese constituency. The PNC remained in power with a single leader, Forbes Burnham, until his death in 1985. The Burnham era of so-called cooperative socialism deteriorated to the point of being maintained by electoral fraud and political violence, creating a de facto one-party state. By the 1980s political closure, suppression of civil society, underdevelopment of the private sector, cooption of the labor and women's movements, and political rent seeking on a grand scale were being translated into economic crisis and mass emigration. This culminated in the withdrawal of World Bank/International Monetary Fund recognition in 1982. Over this period poverty and vulnerability became increasingly widespread, and risk grew with a reactive strategy in flood control of crisis management. The rural areas, predominantly settled by the Indo-Guyanese, were most at risk, with large areas of cultivable land becoming salinated or eroded. Although shortly after independence a sea wall had been built to protect Georgetown, the city's inhabitants received little more support than those living in rural areas (Sir William Halcrow and Partners PLC, 1994b).

With the death of Burnham in 1985, leadership of the PNC was transferred to Desmond Hoyte. Responding to local and international pressures, Hoyte embarked upon a project of liberalization, privatization, and democratization. Consequently, the International Monetary Fund renegotiated a structural adjustment program (SAP) with Guyana in the late 1980s (Ferguson, 1995). Subsequent free elections held in 1992, 1996, and 2001 have been won by the PPP. However, to a large extent, broad policy from this period to the present day has been dictated by the requirements of the SAP. Throughout the 1990s real earnings declined, hitting hardest those living in the urban centers and the Afro-Guyanese community, many of whom have lost their jobs because of downsizing in the public sector. Consequently, Georgetown has resurfaced as a center of extreme vulnerability to coastal flood hazard.

Associated with the SAP is the creation of a participatory or community-sponsored development program. For local communities it provides a second route of resource distribution in parallel with government ministries and local government. The funds are targeted for projects that have been proposed by officially recognized community groups, which, as a response, have also risen greatly in number since 1990. There is, however, some considerable doubt over the meaningfulness of the participation associated with this program, so that notwithstanding superficial decentralization in decision making, deeper structures of political patronage and information asymmetries continue to influence the distribution of resources between and within communities, and so to affect the production of vulnerability to flood hazard (Pelling, 1998).

Contemporary Vulnerabilities in the Urban Environment

The case study presented below brings together data from two sample areas. Both areas, one urban and one periurban, fall within the hydrological regime of the East Demerara Water Conservancy and are considered to be at high risk from flooding (Sir William Halcrow and Partners PLC, 1994a; Pelling, 1996) (Figure 4.1). The periurban case study of 569 households was drawn from six neighborhoods in a settlement on the coast some 16 kilometers east of the capital, the urban study of 240 households was drawn from four neighborhoods in greater Georgetown. The study sites are not meant to be representative for urban Guyana but rather to indicate the types of vulnerabilities and scope for adaptation that exists on the coast. The urban and periurban settlements were chosen because they exhibit high exposure to flooding, with both sites having experienced recent flooding at the time of the research. Within each study, areas of varying economic status and housing tenure existed: these indicators formed the basis of vulnerability assessments. Individual household heads were selected via a random sample of every third household and interviewed using a questionnaire. Following this survey, a 10% subsample of respondents was interviewed in-depth. Household data was supplemented by key informant interviews with local community group and local government leaders. Quantitative data was analyzed using the Statistical Package for Social Scientists (SPSS Inc., Chicago; www.spss.com). Income, housing tenure, gender of household head, and participation in community-based environmental im-

FIGURE 4.1. Settlements and urbanization in the coastal plain of Guyana. Urban settlements are centered on the capital of Georgetown.

provement initiatives were chosen as indicators for vulnerability. For a fuller description of methods and results, see Pelling (1997, 1998).

Both study sites share broadly similar environmental hazards. Rainfall and land drainage are common sources of flood hazard. There are two wet seasons (May–July and November–January), with mean annual precipitation of 2500 millimeters at Georgetown. Numerous small rivers (discharging 250–1,000 square millimeters/second) enter the coastal plain, at which point many of their waters are captured by the East Demerara Water Conservancy which acts as a flood control reservoir. Overland drainage is by a gravity system of drainage and irrigation (D&I) canals running from the Water Conservancy dam to "clokers" at the sea or river defenses, many of which are only open to discharge for 7–14 hours in 24. Settlements have residential drainage systems integrated with the main drainage canals. The East Demerara Water Conservancy dam and much of the major D&I infrastructure have suffered from inadequate maintenance, resulting in overtopping and the threat of dam breaching (Camacho, 1994). For periurban settlements outside Georgetown, the coast is poorly defended against flooding (World Bank, 1993). Sea defenses are under most pressure between October and March when spring high tides coincide with afternoon onshore winds.

The Periurban Study

Between February 1990 and June 1996 seven periurban floods were reported in the press. In these cases direct economic losses incurred by petty agriculturists were estimated to be around $US1,000–$US50,000 per village (Pelling, 1996). At the time of the study, the most recent flood to impact directly upon the community was a sea wall breach in 1993, 2 years before the survey. This event was linked to two mortalities from typhoid and had resulted in around one-quarter of respondents being forced to give up petty agriculture businesses because of the threat of flooding. With echoes of Chamber's (1995) ratchet effect of vulnerability, this created a vicious circle of reduced livelihood options, increasing household vulnerability and further reducing livelihood sustainability. The locations of the neighborhoods in the periurban study are shown in Figure 4.2.

While high income was a route to reducing flood exposure (indicated by home ownership, secure dwelling construction, low-density household and neighborhood), it also provided the opportunity for greater loss of physical assets. While the difference was not statistically significant,[1] direct economic losses from flood damage were recorded by 82% of households earning more than $G30,000/month,[2] compared to 75% of households earning less than $G30,000/month. Despite greater risk of financial loss, it is likely that higher income households were more able to absorb losses than lower income households.

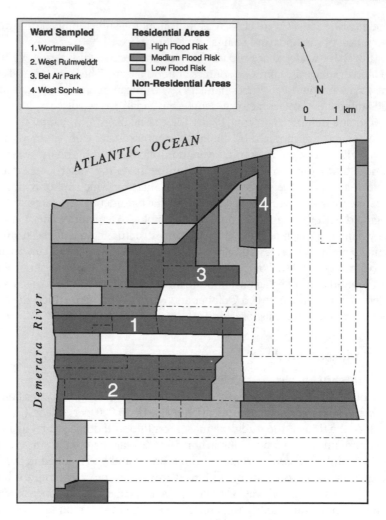

FIGURE 4.2. Periurban neighborhoods in the study sample, outskirts of Georgetown, Guyana.

A clearer indicator of vulnerability was household tenure. Renters, at 64%, were made vulnerable through being unable to adapt properties to reduce exposure to flooding, compared to 80% of owners and 87% of squatters, a statistically significant difference. Renters were also less likely to have invested in communal action to improve the local environment, 9%, compared to 14% of owners and 23% of squatters. Owners had most access to septic tanks, 11%, compared to 4% of renters and 0% of squatters. However, the common use of pit latrines placed the entire population's health at risk during times of flooding. Squatter households were the most vulnerable group, with many squatters having little access to economic resources, living in over-crowded households, and exposed to unsanitary conditions. Some security had been gained by individuals through the raising of squatter dwellings on stilts. Moreover, opportunities existed to mobilize social assets, shown by the high proportion of respondents with relatives also living in the settlement, 72%, compared to 66% of renters and 67% of owners. As Figure 4.2 shows, many squatters also lived in a site away from the sea wall with lower risk of flooding. Hence, locational or physical assets coupled with individual adaptation reduced risk among this potentially highly vulnerable group.

Households in this predominantly Indo-Guyanese periurban village with female or male/joint headship showed very similar patterns of vulnerability, risk, and flood impact. Female-headed households were differentiated only by a lower likelihood of participation in community-based action, 9%, compared to 25% of male/joint households (statistically significant), and by a lower proportion of respondents with relatives living in the village, 66%, compared to 73% of male/joint households, both indicating that social isolation was experienced by this group.

The Georgetown Study

Since no recent change in annual rainfall patterns have been noted (Kemp, 1993, 1994), the observed increase in flooding in Georgetown has been associated with a range of human processes. Impervious areas within Georgetown increased by 50% between 1963 and 1993, raising the volume of runoff channeled through Georgetown's drainage system. At the same time, drainage capacity has been reduced due to the infilling of drains, inadequate maintenance of existing drainage, the use of drains for informal refuse disposal, and the use of drainage reserves for informal housing and petty agriculture. Since 1989 uncontrolled urban expansion into unserviced areas has similarly increased city vulnerability to flooding from high rainfall events (Sir William Halcrow and Partners PLC, 1994b). Sea level rise will further reduce the efficiency of the city's gravity drainage system (Camacho, 1994) and may induce a rise in groundwater level. Climate change adds further uncertainty to hydraulic systems, with global warming being associated with increased precipitation. The locations of the four study areas are shown in Figure 4.3. All sites were classified as being at high risk of flooding (Sir William Halcrow and Partners PLC, 1994a).

Since the 1981 census recorded 174,000 residents in Georgetown, population growth has been slow and at times negative, placing Georgetown in the unusual position of being a city suffering from environmental hazard and economic hardship but without the high rates of population growth with which such conditions are usually associated (Chan & Parker, 1996). In this case it is a

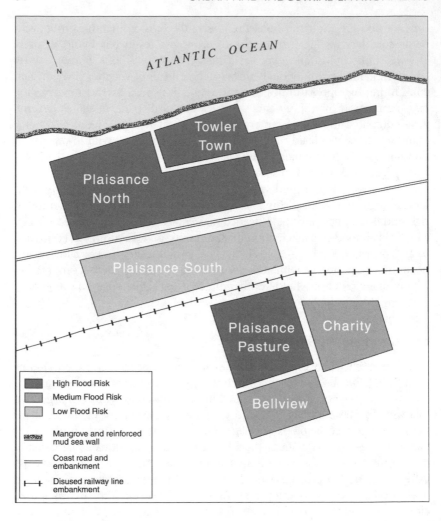

FIGURE 4.3. Urban neighborhoods or wards in the study sample that were evaluated regarding flood risk. The urban neighborhoods are located in Georgetown, Guyana. Redrawn from Sir William Halcrow and Partners PLC (1994b).

failure in service provision rather than increased demand that has led to a worsening of environmental quality and increased environmental hazard. Several localized floods occur in Georgetown each year, with press reports mentioning 21 between January 1990 and June 1996 (Pelling, 1997). It has been estimated that 30% of Georgetown's population suffer flooding at least once a year (Sir William Halcrow and Partners PLC, 1994a). Because flooding was imposed upon a largely preexisting urban morphology, all social classes and urban environments are potentially at risk. The following account followed a storm that released 153 millimeters of rain over 24 hours on 12 November 1995.

A high proportion of low-income households lived in squatter dwellings, 35%, compared to 18% of households with incomes above $G30,000/month; and used pit latrines, 47%, compared to 22% of households with incomes above $G30,000/month; they were more often headed by women, 35%, compared to 16% of households with incomes above $G30,000/month; but more active in community-based action, 34%, compared with 25% of households with incomes above $G30,0001 month than higher income households. This group recorded a high proportion of flooded dwellings, 30%, compared to 21% of households with incomes above $G30,000/month.[3]

Household tenure revealed a continuum of increasing vulnerability from owners through renters to squatters based upon reported access to economic assets and physical infrastructure. However, the urban squatter group reduced exposure by engaging in household adaptation, 66%, compared to 48% of owners and 28% of renters; and by participation in community action to enhance local environmental conditions, 49%, compared to 20% of renters and 7% of owners (all statistically significant differences). It is perhaps partly because of communal action that the squatter community included in this study did not flood on 12 November. There was little difference between renters and owners in the likelihood of flooding or of suffering direct economic losses or health impacts.

Female-headed households were marked by low access to economic resources: only 25% had incomes above $G30,000/month, compared to 48% for male-/joint-headed households. Rental accommodation was the most common form of tenure, which in part explains why individual dwelling adaptation was low: 33% had yard or dwellings raised, compared to 43% of male-/joint-headed households. In contrast to the periurban sample, urban women-headed households were active in community-based action, 20%, compared to 13% of male/joint households. Although not statistically significant, this difference indicates a trend.

Further analysis at the neighborhood level found that high-income households (e.g., in Bel Air Park) were exposed to flooding but had been able to manage vulnerabilities through the transfer of flood impacts from health to economic investment and loss. Similar avoidance of flood impacts (though from a very vulnerable base) was identified for low-income households living in recently constructed self-help dwellings (West Sophia) who had responded to contemporary environmental conditions and so reduced individual vulnerabilities. Such flexibility was not encountered in formal housing areas where dwelling form and drainage infrastructure were more fixed and responsibility for living environments was less clear, being shared between house owners, landlords, tenants, and municipal authorities.

Power, Resistance, and Vulnerability

The preceding evidence has shown the influence of inequality in access to political, economic, social, and environmental assets on shaping the geography of
vulnerability to environmental risk in urban Guyana. This section examines
the sociopolitical mechanisms in Guyanese society that allow such inequalities
to persist in the face of disaster losses. In particular, I focus on the weaknesses
of a participatory approach that has become dominant in development discourse and policy in Guyana and in many other countries of the global North
and South.

For a democratic polity to function meaningfully, it requires political
competition and popular participation. Neither of these conditions were met
in the Guyanese case. Despite apparently far-reaching political and economic
reforms made in Guyana since the 1980s, decision-making power continues to
be held centrally by national political elites drawing upon racially defined constituencies. The racial bias in voting patterns undermines meaningful political
competition. Similarly, while popular participation is often lauded in policy
documentation and political rhetoric, it is less easy to identify in the mechanisms of everyday decision making.

In using a political ecology framework it is important to understand the
interaction of key political actors and to identify the pressures that shape such
interactions. In both case studies, community-based organizations reflect the
partisan political landscape of Guyana. Horizontal linkages between grassroots actors remain undeveloped, and may even have been weakened by recent
participatory projects: vertical linkages are underdeveloped or have been incorporated into systems of patronage. Guyana's political legacy made acceptable the capture of community leadership by local political elites in the
periurban case, and explains the preference expressed, in particular by squatter groups, in Georgetown to make alliances with the dominant political party
of the day. There remains a gap between the lowest levels of government (municipal, local) and host communities. Similarly, in Georgetown, there is a gap
between the municipal authority and the national government. The size of
these gaps allows inefficient governance to operate, and opens resource distribution for environmental management to rent seeking and partisan political
influence (van der Linden, 1997). This shapes the distribution of environmental infrastructure and services and allows political power to produce distorted
geographies of neighborhood vulnerability.

To assess how local power structures are shaped, we need to examine the
legitimacy and roles of the leaders and members of neighborhoods. The participatory model supported by international donor agencies is principally applied in periurban settlements. Participation in decision making through this
model is undermined from initiation by the top-down construction of community groups in which leaders are self-selecting from within established elite

groups. Community leaders are closely associated with local governments, allowing the cooption of community organization by political elites at the local level. Superficially this is advantageous, providing a tie-in to party patronage networks. At a deeper level, such ties undermine the self-reliance of communities and reinforce the racial and class divisions of Guyanese society. Those community leaders who sought to distance themselves from political partisanship found themselves in competition with local political activists for local support. The majority of respondents acknowledge that the aims of local leaders dominate the wider interests of the host community, and that the community advocate role of leaders is often left unfulfilled. A prime example of this conflict of interest is seen in one periurban leader whose first project was to resurface an access road to his own property and workplace. However, for communities with insufficient economic resources to access goods and services from the private sector, leaders with political connections are likely to be the best means of accessing external support of any kind, while minimizing risk taking and resource expenditure for the individual household. As has been found elsewhere (Desai, 1995), patronage can be a useful resource for marginalized and otherwise excluded groups to access decision-making power.

A common proposition by community development advocates is that the most powerless and at risk in society can improve their position through working together (e.g., Kaufman & Alfonso, 1997). This was not found to be the case in urban Guyana. It was in the most highly vulnerable neighborhoods (with low incomes, agricultural livelihoods, rental and squatter households, inadequate physical infrastructure) that leadership from community-based organizations or patronage-based local authority representation was most lacking. In periurban neighborhoods with formally recognized community organizations, vulnerable individuals (low-income householders, renters, petty agriculturalists, female-headed households, the young and the old) were excluded from decision making, which was the domain of house-owning businessmen with relatively high socioeconomic status. The most vulnerable communities were not served by this system. Neither the self-help nor the empowerment that might have provided bases for social development and the strengthening of local social capital as precursors for more representative community organization were apparent, and consequently an effective mechanism for the distribution of development resources to reduce vulnerability was lost.

In the Georgetown study, structural resource scarcity and inefficient management in the public sector undermined efforts to bring decision making closer to the local level (decentralization and democratization), and the historic marginalization and ongoing underdevelopment of the private sector and civil society prevented a mixed framework for resource distribution of public, private, and civil society mechanisms from evolving. Here, politicization was not only a characteristic of emerging grassroots organization, but also of relations between different scales of government bureaucracy. In this sense, flooding and vulnerability were outcomes of a political discourse as well as of environmental change.

Scott (1985) has noted the weapons of the weak used to resist more powerful members of society. Yet public displays of resistance to the institutional structures that constrained local participation and access to resources were uncommon in Guyana. In Plaisance North, an independent community group sought to dislodge the imposed community group controlled by a senior member of local government, and in Plaisance Pasture a letter to the national press openly criticized the disbursement of community funds within the local authority area. Resistance to structures of political authority can also be read into the high nonpayment of local government property taxes, which reached 33% in Georgetown in 1992, and also in the reluctance of grassroots actors to volunteer labor or resources for participatory projects. The contradictory relationship between leaders and community is best shown here, with one leader responding to low community participation rates by suggesting that local environmental rehabilitation schemes would be made more effective if the requirement for active community support for projects was dropped from funding agency requirements.

More common than public resistance has been withdrawal from community action and a retreat into private adaptation measures within the household or family. Preference for household-centered adaptation (raising yards or modifying dwellings) over communal action (drain cleaning, garbage collection) was commonplace, despite the potentially greater security gains to be made from communal action. Withdrawal needs to be seen in the context of recent Guyanese political history, which promoted the dependency of grassroots actors on external agents for local environmental decision making. It also reflects risk aversion and a wariness of, and uncertainty about, the social and economic costs and benefits that participation might bring to an individual or community.

When we ask why it is that environmental risk persists in cities, it is worth considering who benefits from risk. In this case study flooding and the mitigation resources it attracted were a potential source of income for organizations at the national and the local level, as well as for individuals and private-sector entrepreneurs. Among the private sector many petty or informal workers occupied the service provision roles abdicated by the state, and benefited financially as laborers contracted to raise yards, clear drains, collect garbage, or sell drinking water. However, the underdevelopment of the private sector meant that most entrepreneurial activity in this sector was petty, so that in contrast to the water speculators of Guayaquil (Swyngedouw, 1995), the gains made were modest and not sufficient to raise these individuals out of poverty. Rather, it was the local and national political and economic elites that had the most to gain from the present institutional framework for resource distribution.

The political discourses surrounding urban risk had many parallels with earlier responses to rural environmental risk where, for example, it was at the grassroots level, in this case peasant farmers, that blame was first placed for soil loss in the Sahel. In Guyana, despite a rhetoric of participation, managerialist perspectives of social–environmental relations dominated decision-making discourse. This was seen in the blaming of popular, informal social institutions (cultures, norms, and practices) for local, though widespread, environmental degradation. More recent work (e.g., Leach & Mearns, 1996) has reconceptualized environmental change in rural contexts, placing more emphasis on the reasons for practices resulting in soil loss, and indeed questioning that soil loss has been occurring at all. That the practices of the impoverished can be straightforwardly linked to degraded environments (in which the poor must live) continues to linger in constructions of urban socioenvironmental relations. Here, bad citizens are blamed directly for dumping of garbage and waste into drainage canals, for failing to enter into the public spirit of municipal city cleaning days, of contaminating canals with human and household waste, of allowing cattle to roam the city, and for creating their own vulnerability (and societal costs through health care or eviction expenditure) by colonizing sea wall or canal reserves for informal housing. In this way the structural problems underlying individual acts are overlooked and proximate causes of vulnerability and risk too easily become the core concern of management discourse.

The political ecology approach draws analysis of urban risk into a historical perspective and into an exploration of the processes and forces that shape human behavior. This is a departure from established disasters studies work, which has tended to look retrospectively at social relations in the context of a single, apparently exceptional, disaster event. But disasters are situated in evolving social contexts. The growing momentum and reach of socioeconomic change associated with economic and political forms of globalization and the local impacts of global environmental change make it difficult to base policy decisions for future development on an analysis of past events. Social and environmental forces, structures, and powers change over time, smoothly or abruptly, and a deeper understanding of the causes of urban environmental risk and human vulnerability calls for an awareness of the trajectories of change and interrelationships between co-evolving human and environmental elements of the whole.

CONCLUSION

Political ecology puts explanatory emphasis on political power and social organization in the shaping of the "natural" environment, and encourages a historical examination of the processes that produce geographies of environmental and social distress. At first glance the built environment appears to fall outside of such a conceptualization, with natural processes being apparently subjugated by human agency. However, the domination of nature, even in cities, is only partial. Cities are constrained, for example, by their lack of capacity to manage water and air resources and to cope with the production of waste. At times, and in places of social vulnerability, these constraints can manifest as vulnerability to environmental hazard.

The development history of Guyana shows the influence of local agency, as well as of dynamic pressures in the core regions of the global political economy, for shaping society–nature relations and experiences of environmental hazard. Within urban Guyana, these pressures manifest as local conditions that characterize vulnerability to flooding. They include reduced access to economic assets for the poor majority, inadequacy in infrastructure provision, the gendered and ethnic nature of social systems, partisan politics, and an underdeveloped civil society. Those social groups most at risk were low-income households, renters, squatters, female-headed households, the young, the old, and the infirm. However, between urban and periurban households, vulnerability was found to be differentially constructed according to livelihood; the cultural norms of different ethnic groups, especially in relation to female participation in the labor market and community activities: and the mechanisms of infrastructure provision. Consequently in the urban sample it was inner-city residents who endured the greatest vulnerability, while in the periurban sample petty agriculturalists were most at risk.

The status of local constructive social capital (social networks underpinned by cultural norms that promote horizontal integration and inclusivity with accountability to the wider community) continued to be deeply influenced by the legacy of the centralized and oppressive political regime of the 1970s and 1980s, which eroded civil society to its most basic elements of social organization (the household and the family). The suppression of the private sector and civil society over this period put the public sector and hence political decision makers at center stage in the distribution of resources for environmental management. Restructuring decision-making networks to enhance grassroots participation has, therefore, to overcome both a withdrawal of civil society and an entrenched (and adaptive) political elite. With this in mind, it is not surprising that the superficial and quick-fix approaches to the construction of beneficial social capital that have been employed in the present participatory mechanisms have failed to make decision making more inclusive, to empower grassroots actors, or to strengthen horizontal social linkages between grassroots actors.

The political ecology framework has allowed the socioeconomic and political forces shaping decision-making structures to emerge as important elements in the distribution of assets, and hence in the production of geographies of vulnerability in urban Guyana. These forces operate at local, national, and international levels, each affecting the others and pointing toward specific areas for policy reform to alleviate vulnerability. At the local level, the reluctance of individuals to become involved in public life and the contradictory nature of leadership undermine attempts to bring grassroots actors into the decision-making circle. Addressing these characteristics of Guyanese society is likely to be a long-term project. People can not be coerced into participating and will only do so when they see little risk for themselves and have the confidence to expect a good chance of gaining environmental improvement through their actions. Changing the culture of local leadership in Guyana is similarly a long-term project, although some steps in this direction have been taken with international nongovernmental organizations providing leadership training courses. Local-level challenges are reflected in national society, with political leadership failing to escape from the short-termism of racial politics. But ongoing discussions of power sharing offer some hope of a way forward. Internationally, the influence of international financial institutions has grown with time. The World Bank, the International Monetary Fund, and the Caribbean Development Bank have been instrumental in pushing for national policies of privatization, democratization, and participation. The speed of these reforms has tended to limit their positive impact. Privatization has created urban unemployment and a large informal sector, democratization has promised representation without it being delivered, and participation has been captured by local elites.

Discussing urban risk in terms of democracy, participation, leadership, and grassroots resistance is in stark contrast with the dominant approach to hazard management, which concentrates on physical forces and engineering solutions. The political ecology framework should not be seen as an attempt to replace the established management paradigm, but rather to add a further dimension to our understanding of the production of risk that highlights the need to view vulnerability as being deeply embedded in ongoing development policies and discourse.

ACKNOWLEDGMENTS

Funding for this research was provided by the Global Environmental Change Programme of the Economic and Social Research Council.

NOTES

1. In this chapter, all tests for significant difference between two distributions used the binomial test.
2. $G150 = $US1 (1999).
3. All differences between low-income and high-income respondents are significant.

REFERENCES

Agyeman, J. (2000). *Environmental justice: From the margins to the mainstream.* London: Town and Country Planning Association.

Blaikie, P. M., Cannon, T., Davis, I., & Wisner, B. (1994). *At risk: Natural hazards, peoples, vulnerability and disasters.* London: Routledge.

Bryant, R. L. (2001). Political ecology: A critical agenda for change? In N. Castree & B. Braun (Eds.), *Social nature: Theory, practice and politics* (pp. 151–169). Oxford, UK: Blackwell.

Camacho, R. F. (1994). *Policy and environmental aspects of water control schemes.* Devizes, UK: Camacho Environmental Consultants.

Chambers, R. (1995). Poverty and livelihoods whose reality counts. *Environment and Urbanisation, 7*(1), 173–204.

Chan, N. W., & Parker, D. J. (1996). Response to dynamic flood hazard factors in peninsular Malaysia. *Geographical Journal, 162*(3), 313–325.

Desai, V. (1995). *Community participation and slum housing: A study of Bombay.* London: Sage.

Drèze, J., & Sen, A. (1989). *Hunger and public action.* Oxford, UK: Oxford University Press.

Ferguson, T. (1995). *Structural adjustment and good governance: The case of Guyana.* Georgetown, Guyana: Guyana National Printers.

Garza, G. (1999). Global economy, metropolitan dynamics and urban policies in Mexico. *Cities, 16*(3), 149–170.

Hewitt, K. (Ed.) (1983). *Interpretations of calamity: From the viewpoint of human ecology.* Boston: Allen & Unwin.

Hewitt, K. (1997). *Regions of risk: A geographical introduction to disasters.* Essex, UK: Longman.

Kaufman, M., & Alfonso, H. D. (1997). *Community power and grassroots democracy.* London: Zed Books.

Kemp, S. (1993). *A statistical analysis of Georgetown rainfall: A proportionate approach.* Georgetown, Guyana: Government of Guyana, Ministry of Agriculture.

Kemp, S. (1994, September 12). *Observed climate trends in Guyana.* Paper presented at the Climate Change Conference, Ministry of Agriculture, Georgetown, Guyana.

Leach, M., & Mearns, R. (1996). *The lie of the land.* London: Currey.

Mabey, N., & McNally, R. (1999). *Foreign direct investment and the environment: From pollution havens to sustainable development.* London: Worldwide Fund for Nature–United Kingdom.

McGranahan, G., Jacobi, P., Songsore, J., Surjadi, C., & Kjellén, M. (2001). *The citizens at risk: From urban sanitation to sustainable cities.* London: Earthscan.

Mitchell, J. K. (Ed.). (1999). *Crucibles of hazard: Mega-cities and disasters in transition.* Tokyo: UNU Press.

Moser, C., Gauhurst, M., & Gonan, H. (1994). *Urban poverty research sourcebook: Module 2: Sub-city level research.* Washington, DC: World Bank.

O'Keefe, P., Westgate, K., & Wisner, B. (1976). Taking the naturalness out of natural disasters. *Nature, 260,* 566–567.

Özerdem, A. (2003). Disaster as manifestation of unresolved development challenges:

The Marmara earthquake, Turkey. In M. Pelling (Ed.), *Natural disaster, development and global change*. London: Routledge.

Peet, R., & Watts, M. (Eds.). (1996). *Liberation ecologies: Environment, development, social movements*. London: Routledge.

Pelling, M. (1996). Coastal flood hazard in Guyana: Environmental and economic root causes. *Caribbean Geography, 7*(1), 3–22

Pelling, M. (1997). What determines vulnerability to floods?: A case study in Georgetown, Guyana. *Environment and Urbanisation, 9*(1), 203–226.

Pelling, M. (1998). Participation, social capital and vulnerability to urban flooding in Guyana. *Journal of International Development, 10*, 469–486.

Pelling, M. (2001). Natural disasters? In N. Castree, & B. Braun (Eds.), *Social nature: Theory, practice and politics* (pp. 170–188). Oxford, UK: Blackwell.

Pelling, M. (2003) *The vulnerability of cities: Social adaptation and natural disaster*. London: Earthscan.

Scott, J. C. (1985). *Weapons of the weak: Everyday forms of peasant resistance*. London: Yale University Press.

Sen, A. (1981). *Famines and poverty*. London: Oxford University Press.

Sir William Halcrow and Partners PLC. (1994a). *Sewerage system report, Georgetown water and sewerage master plan* (Part 3, Vol. 1). Burdrop Park, Swindon, UK: Author.

Sir William Halcrow and Partners PLC. (1994b). *Household surveys, Georgetown water and sewerage master plan* (TN/GMPI 09). Burdrop Park, Swindon, UK: Author.

Smith, N. (1984). *Uneven development*. Oxford, UK: Blackwell.

Swift, J. (1989). Why are rural people vulnerable to famine? *IDS Bulletin, 20*(2), 8–15.

Swyngedouw, E. A. (1995). The contradictions of urban water provision: A study of Guayaquil, Ecuador. *Third World Planning Review, 17*, 387–405.

United Nations. (1989). *Prospects for world urbanization*. New York: Author.

van der Linden, J. (1997). On popular participation in a culture of patronage: Patrons and grassroots organisations in a sites and services project in Hyderbad. *Environment and Urbanisation, 9*(1), 81–90.

White, G. F., Kates, R. W., & Burton, I. (2002). Knowing better and losing even more: The use of knowledge in hazards management. *Environmental Hazards, 3*(3–4), 81–92.

Wildavsky, A. (1988). *Searching for safety*. London: Transaction Books.

World Bank. (1993). *Guyana: Public sector review*. Washington, DC: Author.

CHAPTER 5

Modernity and the Production of the Spanish Waterscape, 1890–1930

Erik Swyngedouw

Spain is arguably the European country where the water crisis has become most acute in recent years. Since 1975, demand for water has systematically outstripped supply and, despite major and unsustainable attempts to increase pumping of groundwater and to develop a more intensive use of surface water, the problem has intensified significantly. The recent 1991–1995 drought, which affected most of central and southern Spain, prompted intense political debate, particularly as the cyclical resurgence of diminished water supply from rainwater coincided with the preparation of the recently (1999) approved Second National Hydrological Plan and the implementation of a new water law (Gómez Mendoza & del Moral Ituarte, 1995; del Moral Ituarte, 1996).

The political and ecological importance of water is not, however, only a recent development in Spain. Throughout the 20th century, water politics, economics, culture, and engineering have infused and embodied the myriad tensions and conflicts that drove and still drive Spanish society. And although the significance of water on the Iberian Peninsula has attracted considerable scholarly and other attention, the combined socioenvironmental nature has

Adapted by Erik Swyngedouw from the article "Modernity and Hybridity: Nature, *Regenerationismo*, and Production of the Spanish Waterscape, 1890–1930," published in *Annals of the Association of American Geographers, 89*(3), 443–465 (1999). Adapted by permisison of Blackwell Publishing. The adapting author is solely responsible for all changes in substance, context, and emphasis.

remained largely unexplored. Studies of water and water resources have tended to cleave the role of water politics, water culture, and water engineering in shaping Spanish society, on the one hand, and the contemporary water geography and ecology of Spain as the product of centuries of socioecological interaction, on the other hand. The hybrid character of the water landscape, or "waterscape," comes to the fore in Spain in a clear and unambiguous manner. Hardly any river basin, hydrological cycle, or water flow has not been subjected to some form of human intervention or use; not a single form of social change can be understood without simultaneously addressing and understanding the transformations of and in the hydrological process.

I intend to situate the political–ecological processes around water in Spain in the context of what Neil Smith (1984) defined as "the production of nature." In particular, I shall argue that the tumultuous process of modernization in Spain and the contemporary condition, both in environmental and political–economic terms, is wrought from historical spatial–ecological transformations. Modernization in Spain was a decidedly geographical project and became expressed in and through the intense spatial transformation of Spain in the 20th century. This transformation is one in which water and the waterscape play a pivotal role. The contradictions and tensions inherent in the process that is commonly referred to as "modernization" are, I maintain, expressed by and worked through the transformation of nature and society. The "modern" environment and waterscape in Spain is what Latour would refer to as a "hybrid," a thing-like appearance (a "permanence" as Harvey, 1996, would call it) that is part natural and part social, and that embodies a multiplicity of historical–geographical relations and processes.

I would argue, with Latour (1993), that the modern tactic of separating and purifying things natural from those social resides in the conceptual and discursive construction of the world into two separate, but profoundly interrelated, realms—nature and society—between which a dialectical relationship unfolds. The debate, then, becomes a dispute about the characteristics of this dialectical relationship, its implications, and the absence/presence of an ontological foundation from which nature and the social are distinguished and distinguishable. This form of dialectical argumentation runs as follows: Humans encounter nature with its internal dynamics, principles, and laws as a society with its own organizing principles. This encounter inflicts consequences on both. The dialectic between nature and society becomes an external one, that is, a recursive relationship between two separate fields, nature and society, which is mediated by material, ideological, and representational practices. The product, then, is the thing (object or subject) that is produced out of this dynamic encounter.

Neil Smith (1984, 1996), in contrast, insists that nature is an integral part of a process of production or, in other words, society and nature are integral to each other and produce in their unity permanencies (or thing-like moments).

The notion of "the production of nature," borrowed and reinterpreted from Lefebvre (1991), suggests that socionature itself is a historical–geographical process (i.e., time- and place-specific). It insists on the inseparability of society and nature and maintains the unity of socionature as a process. In brief, both society and nature are produced, and hence are malleable, transformable, and potentially transgressive. Smith does not suggest that all nonhuman processes are socially produced, although he insists that all nature, including social nature, is a historical–geographical process (see also Levins & Lewontin, 1985; Lewontin, 1993). He argues instead that the idea of some sort of pristine nature ("First Nature" in Lefebvre's account) becomes increasingly problematic as new "natures" (in the sense of different forms of "nature") are produced over space and time. It is this historical–geographical process that led Haraway and Latour to argue that the number of hybrids and quasi-objects proliferates and multiplies. Indeed, from the very beginning of human history, but accelerating as the modernization process intensified, the objects and subjects of daily life became increasingly more socionatural.

Consider, as examples of this intensification, the socioecological transformations of entire ecological systems (e.g., through agriculture), the sand and clay metabolized into concrete buildings through the labor process, the acceleration and ever-expanding socioecological footprint of the urbanization process, or the contested production of new genomes (such as OncoMouse^{TM}; Haraway, 1997). Of course, the production process of socionature embodies both material processes and the proliferating discursive and symbolic representations of nature. As Lefebvre (1991) insisted, the production of nature transcends material conditions and processes; it is also related to the production of discourses of nature (by scientists, engineers, etc.), on the one hand, and to powerful images, symbols, and discourses on nature (virginity, a moral code, originality, survival of the fittest, wilderness, etc.) through which Nature becomes represented, on the other hand.

Therefore, if we maintain a view of dialectics as internal relations (Olman, 1993; Balibar, 1995; Harvey, 1996) as opposed to external recursive relationships, then we must insist on the need to transcend the binary formations of nature and society and develop a new language that maintains the dialectical unity of the process of change as embodied in the thing itself. "Things" are hybrids or quasi-objects (subjects and objects, material and discursive, natural and social) from the very beginning. By this I mean that the "world" is a process of perpetual metabolism in which social and natural processes combine in a historical–geographical production process of socionature, whose outcome (historical nature) embodies chemical, physical, social, economic, political, and cultural processes in highly contradictory but inseparable manners. Every body and every thing is a mediator, a "hybrid," part social, part natural (but without discrete boundaries), which internalizes the multiple contradictory relations that redefine and rework every body and every thing.

My perspective, broadly situated within the political ecology tradition, draws critically from recent work proposed by ecological historians, cultural critics, sociologists of science, critical social theorists, and political economists. Although researchers working within mainstream perspectives pay lip service to considering the hydrological cycle as a complex, multifaceted, and global network, one that includes physical as well as human elements, they rarely overcome the dualisms of the nature/society divide and they continue to isolate parts from the totality (see, for a review, Castree, 1995; Demeritt, 1994; and Gerber, 1997). My main objective is to bring together what has been severed for too long by insisting that nature and society are deeply intertwined. I excavate the origins of Spain's early-20th-century modernization process (1890–1930) as expressed in debates and actions around the hydrological condition. I intend to structure a narrative that weaves water through the network of socionatural relations in ways that permit me to recast modernity as a deeply geographical, although by no means coherent, homogeneous, total, or uncontested, project. If the social and the natural cannot be severed, but are intertwined in perpetually changing ways in the production process both of society and of the physical environment, then the rather opaque idea of "the production of nature" may become clearer. In sum, I seek to document how the socionatural is historically produced to generate a particular, but inherently dynamic, geographical configuration.

THE PRODUCTION OF NATURE:
WATER AND MODERNIZATION IN SPAIN

The production of the modern Spanish waterscape started at the turn of the 20th century when a distinct discourse and rhetoric of modernization emerged. This modernization drive, which permeated the whole of Spanish society, would generate the anchoring framework for key social, political, cultural, and technical debates and practices until the present day. The modernizing desires of broad strata of Spanish society attempted to construct hegemonic, and apparently socially and politically progressive, visions through the social production of nature. The dialectics of modernization as expressed in Spain's hydropolitics will be documented with an eye toward identifying the relationship between the process of producing a "new" nature and the ebbs and flows of dominant political–economic relations. Multiple narratives will be woven together to reconstruct the relations of power inscribed in the iscursive, ideological, cultural, material, and scientific practices through which the Spanish waterscape became constructed and reconstructed as a socionatural space that reflects Spain's contested modernization process and the relations of power inscribed therein.

The history of Spain's modernization has been a history of altering, redefining, and transforming the very physical characteristics of its waterscape, a process that accelerated from the late 19th century onward. At that time, Spain—belatedly, somewhat reluctantly, and almost desperately—launched itself on a path of accelerating modernization. Today, the country has almost 900 dams, more than 800 of which were constructed in the 20th century (see Figure 5.1). Not a single river basin has not been altered, managed, engineered, and transformed. Water has been an obsessive theme in Spain's national life for more than 100 years and the quest for water continues unabated (del Moral Ituarte, 1998). From the turn of the 20th century onward, water rapidly became a prime consideration in national political, socioeconomic, and cultural debates. Under Franco, the great expansion of hydraulic infrastructures reshaped the hydraulic geography of Spain in fundamental ways. Present discussions in Spain over the need for a new National Hydrological Plan, the debates over the introduction of a new water law, and the relentless demands of cities, regions, and industries for ever more water are testimony to a continuing and ever intensifying conflict and struggle over the trajectory of Spain's modernization process.

Modernization as a Geographical Project: The Production of Space/Nature

While other European imperialist countries were consolidating their geographical expansion overseas at the end of the 19th century, the traditional Spanish elites found themselves in a highly traumatic condition with the loss in 1898 of their last colonial possessions—Cuba, Puerto Rica, and the Philippines—after a disastrous "War of Independence" (Carr, 1983; Figuero & Santa Cecilia, 1998; Fusi & Palafox, 1998). Faced with a mounting economic crisis, growing social tensions, a rising bourgeoisie in the north, and an antiquated and still largely feudal social order in the south that was lamenting the military defeat, Spanish progressive cultural, professional, political, and intellectual elites were desperately searching for a way to revive or to "regenerate" the nation's social and economic base. This drive to revive the nation's "spirit" became known as *"el regenerationismo"* (Fusi & Palafox, 1998). Emerging from growing discontent from the 1870s onward, regenerationism became associated with a movement, the "Generation of '98," a loose group of intellectuals and modernizing elites who were particularly concerned with reviving and modernizing Spain in the context of the twin drama of internal disintegration and the loss of external imperial power (Figuero & Santa Cecilia, 1998).

In the absence of an external geographical project as the foundation for modernization, the Spanish modernizing elites concentrated on a national program that would be equally geographical but founded on the radical transformation of *Spain's* geography, and in particular, its water resources (Gómez Mendoza & Ortega Cantero, 1987). This concern is voiced, among others, by

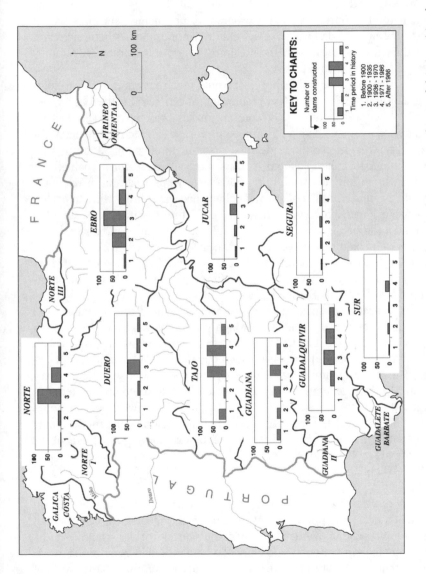

FIGURE 5.1. The Iberian Peninsula, Spain, and the evolution of dam construction for each of the Hydrographic Confederations (river basin authorities). *Source:* Ministerio de Obras Públicas y Urbanismo (1990) *Plan Hidrológico—Síntesis de la Documentación Básica* (Madrid: Dirección General de Obras Hidraulicas), pp. 32–33.

Lucas Mallada (1841–1921), an engineer and geologist, who lamented the fate of Spain and the "causes of the poverty of Spain's soil" (Mallada, 1882; see also Ayala-Carcedo & Driever, 1998) and advocated a program of reworking Spain's natural resource basis. This program of producing new space within the country embodied physical, social, cultural, moral, and aesthetic elements, fusing them around the dominant and almost hegemonic ideology of national development, revival, and progress.

This national geographical project would revolve around the hydrological/agricultural nexus. Spain's "geographical problem" became the axis around which both the sociocultural and economic malaise was explained, and where the course of action resided. It permitted progressive elites to raise social problems (class struggle, economic decline, mass unemployment) as important issues without formulating them in class terms. This, in turn, enabled the formation of an initially weak, but gradually growing, coalition of reformist socialists, populists, industrialists, and enlightened agricultural elites into a hegemonic block with a modernist vision of Spain's future—an alliance aimed to defeat the traditionalists *and* to keep revolutionary socialists and anarchists at bay. Although coalitions, objectives, and means would change over time, the geographical basis for modernization would remain the guiding principle for this hegemonic vision that would become the pivot of Spain's development until the end of the fascist Franco regime.

The realization of such an ambitious project of mobilizing resources and educating the people demanded thorough geographical knowledge, though this was knowledge of a particular kind. In this view, the only means by which to solve the national problem was through the problem of the land: the physical nature of the territory. Even by 1930, this vision was still the primary leitmotiv for the regenerationist agenda, which, in fact, would only materialize on a grand scale after the end of the Civil War in 1938. A newly published journal in 1930 recapitulated the great geographical mission of the modernizing agenda with the same vigor and passion: "There is nothing more urgent for our national reconstitution than a profound study of our geography and our soil. This will be the seed for the great political rebirth of Spain" (N, 1930, pp. 29–30).

This project to remake Spanish geography as a part of modernization combined a decidedly political strategy, a particular ideological vision, a call for a scientific-positivist understanding of the natural world, a scientific–technocratic engineering mission, and a popular base rooted in a traditional peasant rural culture. Plenty of evidence can be found for this in the work of Joaquin Costa and in that of his contemporaries (for reviews, see Perez De La Dehesa, 1966; Orti, 1976). The revolution *in* the state—but certainly not *of* the state—effected through a politics of spatial and environmental transformation, would center around the defense of the small peasant producer-cum-landowner, communal (state) control of water, educational enhancement, technical–scientific control, and the leap to power of an alliance of small-holders and the new bourgeoisie that hitherto had been largely marginalized by the aristocratic landowning elite and their associated administrators in the state apparatus. At the same time, the focus on restoring or, in fact, expanding landownership through "internal colonization" fostered growth in and concentrated the efforts of an "organically" organized state that brought together reformist intellectuals, some worker movements, and the nascent industrial bourgeoisie in a more-or-less coherent vision of reform against the traditionalists (Ortega, 1975). The geographical project became, as such, the glue around which often unlikely partners could coalesce, while excluding both the more radical, left-wing revolutionaries and the "radical" conservatives. Surely, the sublimation of the many tensions and conflicts within this loose alliance of reformists, when accomplished through a focus on reorganizing Spain's hydraulic geography, served the twin purpose of providing a discursive vehicle to ally hitherto excluded social groups without defining the problem purely in class or other conflictual social terms (see Nadal Reimat, 1981).

This organic and anti-revolutionary (in social class terms) reformism in which the state would take center stage to organize the sociospatial transformation would, after the failed attempts to initiate reform during the first few decades of the 20th century, provide a substratum on which the later Falangist, organicist, and fascist ideology would thrive.

Water as the Linchpin to Spain's Modernization Drive

Los Pantanos o la Muerte! (Dams or Death!)
         —PÉREZ (1999, p. 504)

If the "remaking" of Spain's geography became the great modernizing adagio, then water and hydrological engineering were its master tools. The study of geography centered on problems of fertility, both the lack of water and the infertility of the soil. In 1903, Costa wrote that "the greatest obstacle which prevents our country to improve production is the absence of humidity in the soil because of insufficient or absent rainfall" (cited in Ortega, 1975, p. 37). "Rain rushing to the sea and taking part of the soil with it" was to be avoided at all cost, a parliamentary document of 1912 stated, repeating the already century-old claim (which would be heard again during the 1992–1995 drought) that "not a single drop of water should reach the Ocean without paying its obligatory tribute to the earth" (Gómez Mendoza & Ortega Cantero, 1992, p. 174). Indeed, the dominant view at the time was that "Spain would never be rich as long as its rivers flowed into the sea" (Maluquer de Motes, 1983, p. 96). The regenerationist rhetoric assigned great symbolic value to the often-repeated image of the "mutilating loss of the soil of the Fatherland as a consequence of the 'nature' of the pluvio–fluvial regime" (Gómez Mendoza, 1992, p. 240). In addition, both the modernization and urbanization of industry and the mechanization of agriculture generated "water fever" (Maluquer de Motes, 1983, p. 84).

At base, the hydraulic foundation necessitating *el regenerationismo* resided in the uneven distribution of rainfall, and the torrential and intermittent nature of Spain's fluvial system, which was said to make the country "the antechamber of Africa" (de Reparez, 1906). The great modernizing drive of the revivalists therefore demanded not only an imitation and use of nature, but its *creation*: "[increasing] the amount of fertile soil by making a hydraulic artery system cross the whole country—a national network of dams and irrigation channels" (Gómez Mendoza & Ortega Cantero, 1992, p. 174).

This patriotic mission that required the convergence of all national forces fused around the hydraulic program that became the embodiment and representation of a collective myth of national development. This project was sustained and inspired by *a reformist geographical optimism*, which substituted for the social and political pessimism of Spain's turn-of-the-century condition (Orti, 1984, p. 18). The hydraulic utopia of abundant waters for all would not only produce an "ecological harmony," but would also contribute to the formation of a socially harmonious order. The production of a new hydraulic geography would reconcile the growing social tensions in the Spanish countryside, tensions that were taking acute class forms and resulted from the adverse and conflictual conditions of water scarcity and inequality. According to Alfonso Orti (1984, p. 12), the symbolic power of this material intervention in the production of a new hydraulic geography to achieve "hydraulic regeneration" constituted "a mythical power, a collective illusion and the imagined reconciliation of diverse ideologies." This specific form of regeneration served the productionist logic of the new liberal bourgeoisie that aspired to transform society and space according to the principles of capitalist profitability and aimed at Spain's integration into Europe's modernization process.

This hydraulic regenerationism coincided with an intellectual and professional critical regenerationism, symbolized by the literary *Generación de '98*, which rediscovered, both aesthetically and sociologically, the underdeveloped regions of arid Spain. While symbolic representations of Spain by the traditional elites still reveled in perpetuating the "Leyenda de oro" (see Driever, 1998a), which portrayed Spain as a prosperous, thriving, and successful country, a new generation of essayists, poets, and novelists emerged. Although they were by no means a homogeneous group in intellectual and philosophical terms, they shared a desire for a positivist-scientific representation of the fate of Spain's arid or semi-arid regions and a soul-searching quest to transform and regenerate both "soul and body" (Pérez, 1999, p. 500) of the people (Figuero & Santa Cecilia, 1998). Although normally associated with a small group of intellectuals like Ramiro de Maeztu, Pío Baroja, Miguel de Unamuno, Azorin (José Martínez Ruiz), and Antonio Machado, the influence of the generation of '98 was widely felt and was an integral part of the regenerationist debate and modernizing desires that brewed around the beginning of the 20th century (Figuero, 1998). They lamented the decadence of the political elites and their only perspective for future emancipation resided in a spiritual and political rebirth of the nation and in embracing hydraulic politics.

The fate of the drylands became the symbol of the decline and failure of Spain to modernize. The "hydraulic desire" of the arid lands became the leitmotiv of much of the regenerationist literature at the time. The Generation of '98 often invoked writings of Joaquin Costa or Lucas Mallada to express and represent both the landscape and the sociocultural conditions of dryland Spain. Unamuno, for example, refers to symbolic representations such as "the cruelty of the climate," "the somberness of the landscape," or "the biting and dry soul" to evoke Spain's rural conditions and the characteristics of the people. Ramiro de Maetzu describes his native Castillian landscape as "provinces depopulated like Russian steppes" or "a horrible wasteland inhabited by people whose characteristic quality is their hate of water and trees" (cited in Driever, 1998b, p. 33). Although superficially similar to a crude environmental determinism, their views were rather more inspired by a desire to lift Spain from the doldrums of enduring political malaise and persistent poverty of the masses and to recapture its lost position as a great modernizing nation among the other European states.

The State as Master Socioenvironmental Engineer

Surely such an ambitious perspective to regenerate Spain through a geographical project necessitated concerted action and collective control. The regenerationists welcomed the liberation of international markets and the demise of 19th-century protectionism under which the dryland *latifundistas* of (mainly southern and central) Spain flourished. By 1880, trade liberalization had plunged them into a deep crisis in the aftermath of the expansion of U.S. wheat exports. The traditional landed bourgeoisie was economically weakened as a result, but their political commitment to maintain their power at both national and local scales did not abate. This control permitted the continuation, if not the reenforcement, of a strong protectionist economic policy framework.

Nevertheless, central state intervention to produce a nature amenable to the requirements of a modernized, competitive, and irrigated agriculture was considered essential (Ortega, 1975). The state should intervene directly in the implementation of hydraulic works that would permit it to "remake the geography of the fatherland" (Costa, 1892, cited in del Moral Ituarte, 1998, p. 121), to revive the national economy, and "to regenerate the people" (del Moral Ituarte, 1998, p. 121). The hydraulic politics were for Costa a way to insert Spain within a European socio-spatial framework after its loss of influence in the Americas on the basis of a rural development vision that combined a

Rousseauan ideal with a small-scaled, independent, and democratic peasant society. The promotion of the rural ideal on the basis of a petty bourgeois ideology would become the spinal cord of the liberal state and the route to the Europeanization of the nation (Nadal Reimat, 1981, p. 139).

The growing demand for water and the requirements for a more efficient and equitable distribution of irrigation waters necessitated a fundamental change in the legal status and appropriation rights of water. The liberal revolution in Spain (approximately 1811–1873), which had attempted an institutional (anti-)feudal restructuring to promote capitalist forms of ownership and the circulation of goods as commodities, extended also to what Maluquer de Motes (1983) called the "depatrimonialization of water." Indeed, as with land, water did not have the characteristic of a privately owned and tradable good. The existence of seigniorial rights over water prevented or blocked the development of productive activities that necessitated ever larger quantities of water. The regenerationists turned to the state—after the failure of the Liberal project to defeat the feudal elites—as the agent that could generate a sufficiently large volume of capital to mobilize natural resources. A free-market-based, intensive, and productive national economy, whose accumulation process would accelerate on a par with other northern European states, necessitated a transformation in the state in a double and deeply contradictory sense. First, power relations within the state apparatus needed to change in favor of a more modernist alliance of petty owners, industrialists, and modernizing engineers and, second, the social reform that such a revolution required needed to be supported by the state so that the grand hydraulic works could lay the foundations for a modernizing Spain. Surely, this contradiction turned out to be irreconcilable. These two tasks were of course mutually dependent. Strong traditional forces fought to maintain control over key state functions and prevented the rise to political power of the nascent petty owners and middle classes. This firm hold of the traditional conservative elites blocked most attempts at modernizing the social economy. The mosaic of contradictory forces and the resistance of the traditionalists would stall state-led modernizing efforts, resulting in more acute and openly fought social antagonisms throughout the first two decades of the century. These would eventually pave the way for dictatorial regimes from the 1920s onward.

Despite the collectivist discourse of much of the regenerationist literature, it remained deeply committed to a project of insertion into an international capitalist market. In effect, two models of capitalist accumulation, with evidently different supporting social groups and allies, crystallized around the hydraulic debates at the time. The social, political, and ecological consequences and implications of these two models would differ fundamentally even while sharing an organicist vision of the world. On the one hand, the traditionalists defended a protectionist economic stance and the continuation of existing political and social power relations. On the other hand, the regenerationists advocated a more liberal perspective, a rapid modernization of the economy, and a transformation of sociopolitical power relations. The issue of landownership and the role of water therein revolved around the question of who would own and control what part of the land and its waters. For the regenerationists, for whom petty ownership constituted the way ahead, the hydraulic route was an essential precondition, while the limited possibilities for accumulation pointed to the state as the only body that could generate the required investment funds, on the one hand, and push through the necessary reforms in the face of strong and sustained opposition from the landed aristocracy, on the other hand (Ortega, 1975, 1992). At the same time, the very support of at least some sections of the old elites could be secured via this reformist route, since it did not threaten their fundamental rights as landowners and it defended rural power against the rising tide of the urban industrial elites and proletariat. This was indeed quite central to forging the support of the dominant Catholic groups that defended a solidaristic and organic model of social cohesion.

Purification and the Transformation of Nature: Hydraulic Engineers as Producers of Socionature

The hydraulic intervention to create a waterscape supportive of the modernizing desires of the revivalists without questioning the social and political foundations of the existing class structure and social order was very much based on a respect for "natural" laws and conditions. The latter were assumed to be or thought of as intrinsically stable, balanced, equitable, and harmonious. The hydraulic engineering mission consisted primarily in "restoring" the "perturbed" equilibrium of the erratic hydrological cycles in Spain. Of course, this endeavor required a significant scientific and engineering enterprise, first, in terms of understanding and analyzing nature's "laws," and, second, in using these insights to work toward a restoration of the "innate" harmonious development of nature. The moral, economic, and cultural "disorder" and "imbalances" of the country at the time paralleled the "disorder" in Spain's erratic hydraulic geography, both of which needed to be restored and rebalanced (as nature's innate laws suggested) to produce a socially harmonious development. Two threads have to be woven together in this context: on the one hand, the pivotal position of a particular group of scientists in the hydraulic arena, the Corps of Engineers (Villaneuva Larraya, 1991), and the changing visions concerning the scientific management of the terrestrial part of they hydrological cycle, on the other. Both, in turn, were linked to the rising prominence of hydraulic issues on the sociopolitical agenda at the turn of the century.

The Corps of Engineers, founded in 1799, was (and remains) the professional collective responsible for the development and implementation of public works. It is a highly elitist, intellectualist, "high-cultured," male-dominated, and socially homogeneous and exclusive corporatist organization that has over the centuries taken a leading role in Spanish politics and development (Mateu Bellés, 1995). The decision-making structure is hierarchical and all key managerial and institutional bodies, such as the Junta Consultiva de las Obras Públicas, the hydrological divisions, the provincial headquarters, and ad hoc study commissions, are exclusively "manned" by engineers.

In line with the emergent scientific discourse on orography (i.e., the branch of geography that addresses mountains) and river basin structure and dynamics, the engineering community argued for the foundation of engineering and managerial intervention on the basis of the "natural" integrated water flow of a water basin, rather than on the basis of historically and socially formed administrative regions (see Figure 5.2). The emergent geographical regionalization overlaid the traditional political–administrative divisions of the country, forcing a reordering of the territory on the basis of the country's orographical (i.e., mountain range) structure. The latter, in turn, was portrayed by the engineers as the crucial planning unit for hydraulic interventions. Cano García (1992, p. 312) succinctly summarizes this scientific perspective:

> To revert to the great orographical delimitation for organizing the division of the land represents a contribution made from within the strict field of our discipline [engineering] and at the same time, at least initially, it shows the abandoning of traditional political divisions and the importance of other perspectives and concepts. (my translation)

This scientific and natural division provided an apparently enduring and universal scale for territorial organization in lieu of the more recent political and historical scales associated with politico-administrative boundaries. As Smith (1969, p. 20) argues, "The identity of the drainage basin seemed to offer a concrete and 'natural' unit which could profitably replace political units as the areal context for geographical study." Brunhes (1920, p. 93) insisted on the water basin as the foundation for the organization of the land since "water is the sovereign wealth of the state and its people." Such a view was widely recounted in Spain, and its arguments were rallied in defense of a new orographic administrative organization of the territory (Figure 5.2).

The history of the delimitation of hydrological divisions is infused with the influence of the regenerationist discourse, on the one hand, and the scientific insights gained from hydrology and orography, on the other. The attempt to "naturalize" political territorial organization was part-and-parcel of a strategy of the modernizers to challenge existing social and political power geometries. The construction of and command over a new territorial scale might permit them to implement their vision and bypass more traditional and reactionary power configurations. The complex history of the formation of river

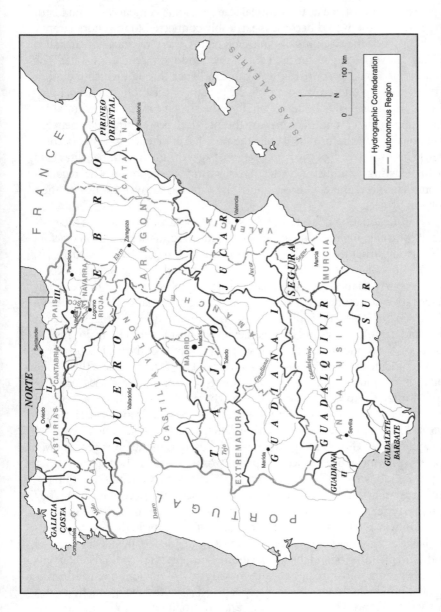

FIGURE 5.2. The river basin authorities of Spain known as "Hydrographic Confederations" (Confederaciones Sindicales Hidrográficas). The entire country of Spain is divided into the territories of these socioenvironmental organizations.

basin authorities and their articulation with other political forms of territorial organization, in particular the national state, is a long, complicated, and tortuous one. "Nature" would become inextricably connected to the choreography of power, while the scientific discourse on nature was strategically marshalled to serve power struggles for the control over and management of water. The river basins would become the scale par excellence through which the modernizers tried to undermine or erode the powers of the more traditional provincial or national state bodies. Therefore, the struggle over the territorial organization of intervention expresses the political power struggles between traditionalists and reformers. While the river basin defenders would become the "founding fathers" of the regenerationist agenda, the traditional elites held to the existing administrative territorial structure of power. The regenerationist engineers thereby incorporated the naturalized river basins into their political project.

This negotiation of scale and the science/politics debate around the scaling of hydraulic intervention and planning raged for almost a century before the current structure was put into place (Cano Carcía, 1992; Mateu Bellés, 1995). The Water Act of 1879 had established that all surface water was common property, managed by the state. This also implied the need to create administrative structures to perform these managerial tasks (Giasante, 1999). The first hydrological divisions (10 in total) were established by royal decree in 1865 and, from the very beginning, they were considered to be major instruments for the economic modernization process. Some of these divisions more or less coincided with major river basins (Ebro, Tajo, Duero), others (like in the south) had a much closer correspondence to provincial boundaries. All were named after the provincial capital city where the head office was located (Mateu Bellés, 1994). Their basic merit was to serve as an institutional basis responsible for the collection of statistical data to assist the study and research of the water cycle. These surveys could then be used as inputs to the real power holders (provincial head offices for public works, special ad hoc commissions, or private industry) (del Moral Ituarte, 1995). This early attempt to set up hydrological divisions came to an end in 1870 when they were abolished. They were party reerected in 1870, reduced to seven, and then abolished again in 1899. Only in 1900, in the wake of the regenerationist spirit, the seven hydraulic divisions were reestablished and their tasks extended to include the detailed study and planning of and the formulation of proposals for hydraulic interventions. However, the ultimate decision-making power would remain at the traditional provincial level, which supervised and executed the hydraulic works, and with the central state for financing and controlling the infrastructure programs (Mateus Bellés, 1994).

It is only from 1926 onward, during the dictatorship of Primo de Rivera, that the current Confederaciones Sindicales Hidrográficas were gradually established as quasi-autonomous organizations in charge of managing water as stipulated by the Water Act of 1879 (Giansante, 1999). The last of these 10 Confederaciones was only finally established in 1961. What had proven impossible to achieve during the first decades of the century was finally implemented during the dictatorship of General Franco. It is also from that moment onward that their names reflected the river basins for which they were responsible rather than the previous political-territorial naming. In addition, they acquired a certain political status with participation from the state, banks, chambers of commerce, provincial authorities, and so on. At each stage the engineers took the lead roles and became the activists of the regenerationist project through the combination of their legitimization of the beholders of scientific knowledge and insights with their privileged position as a political elite corps within the state apparatus. The complex and perpetually changing administrative organization and power structures associated with the successive attempts to establish river basin authorities and their relative lack of power until the 1930s reflect the failure of the early modernizers to fundamentally challenge existing and traditional power lineages and scales (Mateu Bellés, 1994, 1995). It is only from the later 1920s and, in particular, during the Franco era that the regenerationist project was gradually implemented.

CONCLUSIONS

In sum, the regenerationist agenda(s) first maintained that the restoration of wealth in Spain should be based on the knowledge of the laws and balances of nature; second, this restoration required the correction of defects imposed by the geography of the country and particularly "imbalances in its climatic and hydraulic regimes" (Gómez Mendoza & Ortega Cantero, 1992, p. 173); and third, this enterprise of geographical rectification could, because of its range and importance, only be carried out by the central public authorities The hydraulic mission was seen as the solution to the social problems facing Spain at the turn of the century. Failing this, social tensions were bound to intensify, and struggle, if not civil war, would be the likely outcome. Ironically, of course, the voluntarist, powerful, and autocratic hydraulic engineer pursuing a program of imposed reform foreshadowed the fascist (Falangist) ideology. The latter would gain momentum from the early 1920s onward and be consolidated with Franco's fascist victory. The failure of hydraulic politics in the early decades of the 20th century announced what Costa and his literary allies had feared and desperately tried to prevent. Although the debates at the turn of the century indicated a desire to regenerate Spain, conservative forces prevented its actual implementation and social tension intensified, further destabilizing an already highly fragmented and divisive society. The centralizing fascist regimes that emerged from this turmoil could finally push through the production of a new geography, a new nature, and a new waterscape, something the regenerationists of the turn of the century had so desperately advocated, but failed to accomplish.

In this chapter, I have attempted to reconstruct multiple and often contradictory narratives that span a broad range of apparently separate instances such as engineering, politics, economics, culture, science, nature, ideology, and discourse through which the tumultuous reordering of sociophysical space is shaped and transformed, and out of which new socioenvironmental landscapes emerges—landscapes that are simultaneously physical and social, that reflect historical–geographical struggles and social power geometries, and that interiorize the flux and dynamics of sociospatial change. Geographical conditions are reconstructed as the outcome of a process of production in which both nature and society are fused together in a way that render them inseparable, that produces a restless "hybrid" quasi-object in which material, representational, and symbolic practices are welded together. "Doing geography," then, implies the excavation and reconstruction of the contested process of the "production of nature." Of course, this perspective also asks serious questions about who controls, who acts, and who has the power to produce what kind of socionature.

REFERENCES

Ayala-Carcedo, F. J., & Driever, S. L. (Eds.). (1998). *Lucas Mallada—La futura revolución Española y otros escritos Regenerationistas.* Madrid: Editorial Biblotheca Nueva.

Brunhes, J. (1920). *Geographie humaine de la France.* Paris: Hanotaux.

Cano García, G. (1992). Confederaciones Hidrográficas. In A. Gil Olcina & A. Morales Gil (Eds.), *Hitos históricos de los regadíos Españoles* (pp. 309–334). Madrid: Ministerio de Agricultura, Pesca y Alimentación.

Carr, R. (1983). *España: De la Restauración a la Democracia, 1875–1980.* Barcelona: Editorial Ariel.

Castree, N. (1995). The nature of produced nature: Materiality and knowledge construction in Marxism. *Antipode, 27*(1), 12–48.

del Moral Ituarte, L. (1995). El Origen de la Organización Administrativa del Agua y de los Estudios Hidrológicos en España: El caso de la cuence del Guadalquivir. *Estudios Geográficos, 56,* 371–393.

del Moral Ituarte, L. (1996). Sequía y crisis de sostenibilidad del modelo de gestión hidráulica. In M. V. Marzol, P. Dorta, & P. Valladares (Eds.), *Clima y agua—La gestión de un recurso climático* (pp. 179–187). Madrid: La Laguna.

del Moral Ituarte, L. (1998). L'état de la politique hydraulique en Espagne. *Hérodote, 91,* 118–138.

Demeritt, D. (1994). The nature of metaphors in cultural geography and environmental history. *Progress in Human Geography, 18,* 163–185.

de Reparez, G. (1906, 21 July). Hidráulica y dasonomiá. *Diario de Barcelona.*

Driever, S. L. (1998a). "And since Heaven has Filled Spain with Goods and Gifts": Lucas

Mallada, the Regenerationist Movement, and the Spanish environment, 1881–90. *Journal of Historical Geography, 24*(1), 36–52.

Driever, S. L. (1998b). Mallada y el Regenerationismo Español. In F. J. Ayala-Carcedo & S. L. Driever (Eds.), *Lucas Mallada—La futura revolución Española y otros escritos Regenerationistas* (pp. 15–61). Madrid: Biblotheca Nueva.

Figuero, J. (1998). *La España de la rabia y de la idea.* Barcelona: Plaza y Janés Editores.

Figuero, J., & Santa Cecilia, C. G. (1998). *La España des desastre.* Barcelona: Plaza y Janés Editores.

Fusi, J. P., & Palafox, J. (1998). *España 1808–1996: El desafío de la modernidad.* Madrid: Editorial Espasa.

Gerber, J. (1997). Beyond dualism: The social construction of nature and the natural and social construction of human beings. *Progress in Human Geography, 21,* 1–17.

Giasante, C. (1999). *In-depth analysis of relevant stakeholders: Guadalquivir River Basin Authority.* Unpublished manuscript, University of Seville, Department of Geography. (ms. available from author)

Gómez Mendoza, J. (1992). Regenerationismo y regadíos. In A. Gil Olcina & A. Morales Gil (Eds.), *Hitos históricos de los regadíos Españoles* (pp. 231–262). Madrid: Ministerio de Agricultura, Pesca y Alimentación.

Gómez Mendoza, J., & del Moral Ituarte, L. (1995). El plan hidrológico nacional: Criterios y directrices. In A. Gil Olcina & A. Morales Gil (Eds.), *Planificación hidráulica en España* (pp. 331–398). Murcia: Fundación Caja del Mediterráneo.

Gómez Mendoza, J., & Ortega Cantero, N. (1987). Geografía y Regenerationismo en España. *Sistema, 77,* 77–89.

Gómez Mendoza, J., & Ortega Cantero, N. (1992). Interplay of state and local concern in the management of natural resources: Hydraulics and forestry in Spain 1855–1936. *GeoJournal, 26*(2), 173–179.

Haraway, D. (1997). *Modest_Witness@Second_Millennium: FemaleMan©Meets_Onco-Mouse^{TM}.* London: Routledge.

Harvey, D. (1996). *Justice, nature, and the geography of difference.* Oxford, UK: Blackwell.

Latour, B. (1993). *We have never been modern.* London: Harvester Wheatsheaf.

Lefebvre, H. (1991). *The production of space.* Oxford, UK: Blackwell.

Levins, R., & Lewontin, R. (1985). *The dialectical biologist.* Cambridge, MA: Harvard University Press.

Lewontin, R. (1993). *The doctrine of DNA: Biology as ideology.* Harmondsworth, UK: Penguin Books.

Mallada, L. (1882). Causas de la pobreza de nuestro suelo. *Boletín de la Sociedad Geográfica de Madrid, 7*(2), 89–109.

Maluquer de Motes, J. (1983). La despatrimonialización del agua: Movilización de un recurso natural fundamental. *Revista de Historia Económica, 1*(2), 76–96.

Mateu Bellés, J. F. (1994). *Planificación hidráulica de las divisiones hidrológicas 1865–1899.* Unpublished manuscript, University of Valencia, Department of Geography. (ms. available from author)

Mateu Bellés, J. F. (1995). Planificación hidráulica de las divisiones hidrológicas. In A. Gil Olcina & A. Morales Gil (Eds.), *Planificación hidráulica en España* (pp. 69–106). Murcia, Spain: Fundación Caja del Mediterráneo.

N. (1930). Editorial. *Montes e Industrias, 1*(2), 29–30.

Nadal Reimat, E. (1981). El regadío durante la Restauración. *Revista Agricultura y Sociedad, 19*, 129–163.

Olman, B. (1993). *Dialectical investigations*. London: Routledge.

Ortega, N. (1975). *Política agraria y dominación del espacio*. Madrid: Editorial Ayuso.

Ortega, N. (1992). El plan nacional de obras hidráulicas. In A. G. Olcina & A. M. Gil (Eds.), *Hitos historicos de los regadíos Españoles* (pp. 335–364). Madrid: Ministerio de Agricultura, Pesca y Alimentación.

Ortí, A. (1976). Infortunio de Costa y ambigüedad del Costismo: Una reedición acrítica de "Política hidráulica." *Revista Agricultura y Sociedad, 1*, 179–190.

Ortí, A. (1984). Política hidráulica y cuestión social: Orígenes, etapas y significados del regenerationismo hidráulico de Joaquín Costa. *Revista Agricultura y Sociedad, 32*, 11–107.

Pérez, J. (1999). *Historia de España*. Barcelona: Editorial Crítica.

Pérez De La Dehesa, R. (1966). *El pensamiento de Costa y su influencia en el 98*. Madrid: Editorial Sociedad de Estudios y Publicaciones.

Smith, N. (1984). *Uneven development: Nature, capital, and the production of space*. Oxford, UK: Blackwell.

Smith, N. (1996). The production of nature. In G. Robertson, M. Mash, L. Tickner, J. Bird, B. Curtis, & T. Putnam (Eds.), *FutureNatural: Nature/Science/Culture* (pp. 35–54). London: Routledge.

Smith, T. C. (1969). The drainage basin as an historical unit for human activity. In R. J. Chorley (Ed.), *Introduction to geographical hydrology* (pp. 20–29). London: Methuen.

Villanueva Larraya, G. (1991). *La "Politica hidráulica" durante la Restauración 1874–1923*. Madrid: Universidad Nacional de Educación a Distancia.

PART III

ECOLOGICAL ANALYSIS AND THEORY IN RESOURCE MANAGEMENT AND CONSERVATION

CHAPTER 6

The Ivorian Savanna

Global Narratives and Local Knowledge of Environmental Change

Thomas J. Bassett
Koli Bi Zuéli

Although soil degradation, notably accelerated soil erosion, is unquestionably the result of expansion of farming into sensitive areas (*zones sensibles*), little data exist to confirm this fact.

—WORLD BANK (1994, p. 8)

Humanity cannot stand by and let a whole continent disintegrate. . . . The trees, earth, soil, water, wildlife, the urban environment—those are everyone's business.

—FALLOUX AND TALBOT (1993, pp. xiii–xiv)

The image of an entire continent physically disintegrating due to the destructive land use practices of its inhabitants conveys the magnitude of Africa's environmental problems—at least in the eyes of the World Bank. Yet there remains considerable uncertainty about the very processes generating this assumed degradation of the environment. As the World Bank notes for Côte d'Ivoire in the first epigraph, "little data exist to confirm this fact." This is also the case for other parts of Africa (Watts, 1987; Stocking, 1987, 1996). Despite a lack of reliable evidence, the World Bank considers environmental degrada-

Adapted by Thomas J. Bassett and Koli Bi Zuéli from the article "Environmental Discourses and the Ivorian Savanna," published in *Annals of the Association of American Geographers, 90*(1), 67–95 (2000). Adapted by permission of Blackwell Publishing. The adapting authors are solely responsible for all changes in substance, context, and emphasis.

tion to be so widespread that "the business" of environmental planning and regulation is now seen as a global affair, and one that falls within its own purview (Falloux & Talbot, 1993, pp. xiii–xiv). Indeed, since the late 1980s the World Bank has required low-income countries that receive funding from its International Development Association (IDA) to draw up National Environmental Action Plans (NEAPs).[1] It is currently assisting dozens of African governments to develop NEAPs which, in assembly-line fashion, are being produced according to a blueprint (Greve, Lampietti, & Falloux, 1995). NEAPS are heralded as modern vehicles that will lead its member countries down the road to rational and orderly sustainable development. The "fuzzy green" notion of sustainable development is readily adopted by the World Bank because of its compatibility with its basic "technocratic, managerial, and modernist ideology" (Adams, 1995, p. 93).

In this chapter we examine the contents of the NEAP for Côte d'Ivoire and devote special attention to how environmental problems are defined for the northern savanna region. Our goals are threefold. First, we are interested in demonstrating how the Côte d'Ivoire NEAP's desertification narrative of environmental change forms part of what Peet and Watts (1996) call a "regional discursive formation." By this they refer to the highly circumscribed ways in which development problems and solutions are discussed and "which originate in, and display the effects of, certain physical, political–economic, and institutional settings" (Peet & Watts, 1996, p. 16). The environmental discourse of desertification invokes a set of human–geographical processes that typically blame indigenous peoples for not possessing the right technology, knowledge, or land rights systems considered to be important for good stewardship. By defining land degradation in these terms, aid donors and the governments of African states present themselves as possessing the expertise and projects that will stop the desert and create the conditions for "sustainable development." Alternative analyses of environmental change that conflict with or complicate the dominant narrative are typically ignored or dismissed as useless. Only those analyses that "provide the charter or justification for 'development' interventions" are acknowledged as useful by "development" experts (Ferguson, 1990, pp. 68–69).

A second and related goal is to show how geographical scale is central to the World Bank's environmental analyses and interventions. We interpret the World Bank's environmental degradation claims as *continental* in scope and "*everyone's* business" as a scalar politics that legitimates its technical and institutional interventions by defining environmental problems in a way that makes them fall within its discursive frameworks. In this way, "scale [is] not socially or politically neutral but embodies and expresses power relationships" (Swyngedouw, 1997, p. 140).

Our third objective is to contribute to the growing convergence in cultural and political ecology on the need to consider multiple views and employ multiple research methods in assessing and explaining environmental change (Blaikie, 1994, 1995; Batterbury, Forsyth, & Thomson, 1997; see also Turner, Chapter 8, and Zimmerer, Chapter 7, this volume). One of the challenges of doing cultural and political ecology is demonstrating the linkages between a variety of social and bioyphysical processes driving environmental change. As Blaikie (1995, p. 204) has suggested, there is a "plurality of knowledges" on the environment and a host of epistemological and methodological issues that need to be addressed in combining natural and social scientific perspectives. Notwithstanding these formidable challenges, an ecologically invigorated political ecology is emerging that systematically integrates sociocultural and biophysical processes. The emphasis placed in cultural and political ecological research on local knowledge, environmental history, multiscale politics, and socially differentiated resource management practices requires intensive field study and multiple research methods. This chapter presents a methodological route to scholars interested in traversing the intersecting social and biophysical realms of human geography.

ENVIRONMENTAL NARRATIVES
AND REGIONAL DISCURSIVE FORMATIONS

Geographical images and environmental narratives are commonly used as framing devices by developers to set the scene for their subsequent interventions (Mitchell, 1995). In this sense, they form part of the larger conceptual framework of development discourse. Jonathan Crush (1995, p. 3) defines the focus of development discourse analysis as

> the ways that development is written, narrated and spoken, on the vocabularies deployed in development texts to construct the world as an unruly terrain requiring management and intervention; on their stylized and repetitive form and content, their spatial imagery and symbolism, their use (and abuse) of history, their modes of establishing expertise and authority and silencing alternative voices; on the forms of knowledge that development produces and assumes; and on the power relations it underwrites and reproduces.

Images of chaos and environmental devastation run through the Côte d'Ivoire National Environmental Action Plan (NEAP-CI). Land use practices are described as anarchic as peasants and pastoralists are purportedly destroying savanna woodlands through agricultural land clearing, burning, and overgrazing. The plan proposes new laws and land use regulations as a means of controlling what is uncritically described as a crisis situation.

The history of the idea of desertification in West Africa and its centrality to national and global environmental discourses comprises what Peet and Watts (1996, pp. 15–16) theorize as a "regional discursive formation."

> Certain modes of thought, logics, themes, styles of expression, and typical metaphors [which] run through the discursive history of a region, appearing in a variety of forms, disappearing occasionally, only to reappear with even greater intensity in new guises.

The idea and image of an encroaching desert due to overgrazing, deforestation, and especially burning, has expanded and contracted throughout 20th century in West Africa (Fairhead & Leach, 1998; Mortimore, 1998; Swift, 1996). Like Leach and Mearns (1996) and Forsyth (2001), we are interested in examining such narratives for their practical effects in legitimating the interventions of various actors and institutions like the World Bank in rural economic life. For example, the common idea that African land users are recklessly destroying the land due to unregulated land access has been used by a variety of actors in the past and present to meddle in rural land rights systems. The notion that land privatization will motivate people to invest in land conservation and agricultural production is commonly used by government agencies to extend their control over rural peoples and resources (Anderson & Grove, 1987; Bassett, 1993). We argue that global narratives of environmental degradation, despite their disconnect with regional-scale biophysical processes of environmental change, are central to the green neoliberal model linking land privatization with natural resource conservation and agricultural growth.

The following case study of land cover change in the Ivorian savanna at the local and regional scales belies the narrative of environmental devastation contained in NEAP planning documents. Our research shows that desertification is not taking place in northern Côte d'Ivoire. Rather, the most noticeable trend in land cover change is tree and bush encroachment whose dynamics are linked to changing fire regimes and land use patterns with contrasting spatiotemporal rhythms. We suggest that these landscape-level processes fall outside the World Bank's problematic. What the World Bank requires are simple narratives of environmental change. As Ferguson (1990) and Scott (1998) argue, the state and international aid donors seek to simplify the social and ecological worlds of their subjects in ways that make the complex and unwieldy both "legible" and more accessible to their modes of intervention. This process of remaking reality involves the scaling-up of environmental problems to an analytical level at which the development apparatus can intervene most effectively. By defining the "environmental problem" as broad-scale land degradation caused by irresponsible land users who might otherwise be conscientious stewards (if they only possessed land titles!), the World Bank positions itself as possessing the solution: land privatization.

RESEARCH AREA AND METHODS

This study is the product of a collaborative research project carried out by an interdisciplinary group of scholars at the University of Illinois, Urbana–Champaign, and the Institut de Géographie Tropicale (IGT) at the University of Cocody–Abidjan (Côte d'Ivoire). The research focused on the patterns and processes of environmental and social change in two highly contrasting rural communities, Katiali and Tagbanga, in the Korhogo region of northern Côte d'Ivoire. In this chapter, we focus exclusively on the case of Katiali. A community of some 1,900 inhabitants, Katiali is situated 60 kilometers northwest of Korhogo in the subprefecture of Niofouin. Population densities in this area average 16 persons/square kilometer.

Figure 6.1 situates the study area. The land area framed by the box was subject to aerial photo interpretation (see below). Regional rainfall averages 1,000–1,200 millimeters, with most precipitation occurring during the rainy season months of June–October. Soils and vegetation types vary with topography and land use (see Figure 6.3, below).

We collected information on land cover and land use patterns in the Katiali area using a variety of field research and analytical techniques. Survey research methods were used to administer a questionnaire on environmental perceptions to a sample of 38 Senufo and Jula households in the community. Group interviews were also held with Fulβe pastoralists. To assess whether local perceptions of environmental change were congruent with scientific findings, we reviewed the specialist literature on human-induced modifications of savanna vegetation. We then examined aerial photographs located in Côte d'Ivoire and France for different time periods to compare land use/land cover patterns. Geographic information systems (GIS) techniques were utilized to quantify land cover trends. To gain a clearer understanding of vegetation change dynamics on the ground, we inventoried species along 50-meter transects and in 10- × 10-meter plots following the contact point method adapted from Daget and Poissonnet (1971) by César and Zoumana (1995).[2] Finally, we collected environmental policy and planning documents and interviewed individuals involved in the NEAP process in government ministries and at the World Bank's regional headquarters in Abidjan.

Farming and Herding Systems

Rapid changes have occurred in the farming and livestock-raising systems of northern Côte d'Ivoire since the mid-1950s. The case of Katiali is indicative of general trends in the less densely settled areas of the Korhogo region. These include the expansion of cotton from 1% of the cultivated area in the early 1960s to 45% by the late 1980s. Over the same time period there has been a near doubling in the average area cultivated by households. This transformation of ag-

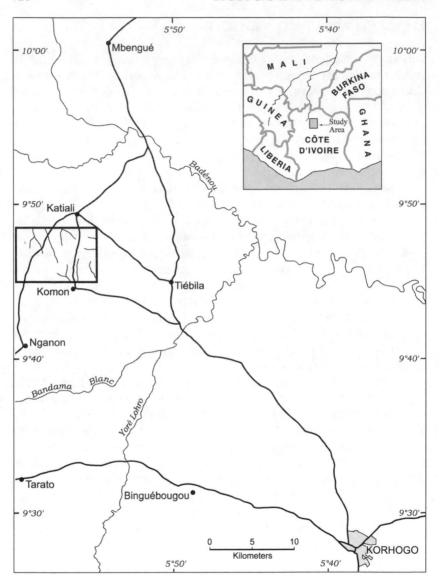

FIGURE 6.1. Location of aerial photo interpretation and vegetation study area, Katiali, Côte d'Ivoire.

ricultural production has been facilitated by the introduction of ox plows into the area. The percentage of Katiali households possessing oxen rose from 37% in 1982 to 92% in 1992. This expansion of ox plows allowed households to increase the area under cultivation from an average of 4.75 hectares to 8.20 hectares. Despite the dominance of cotton in the farming system, peasant farmers continue to cultivate a wide range of food crops.

Katiali's population grew at a low annual growth rate of 0.7% between 1955 and 1984. High infant mortality rates and the emigration of young men and women account for this low rate. Low population densities permit farmers to practice a long-fallow agricultural system. Typically, fields are cultivated for 5–6 years and then put into a 10- to 30-year fallow period. The situation of land abundance and good pastures in the Katiali area has been a magnet to Fulβe pastoralists who have steadily immigrated into the region since the early 1970s (Bassett, 1986). Population growth rates correspondingly increased at an annual rate of 2% between the 1975 and 1988 census.

Parallel to and in many ways associated with the transformation of farming systems, livestock raising has greatly expanded since the early 1960s. In 1962 there were an estimated 115,000 head of cattle owned by Senufo and Jula farmers in the Korhogo region (Société d'Etudes pour le Développement Economique et Sociale [SEDES], 1965, p. 200). Cattle owned by transhumant Fulβe pastoralists who had recently immigrated from the neighboring countries of Mali and Upper Volta (Burkina Faso) numbered no more than 8,000. Thirty years later, there were nearly a million head of cattle concentrated in the northern savanna region. Peasant-owned cattle increased fourfold over this period (1962–1991), while Fulβe cattle grew by a factor of 52. The number of oxen rose from 252 head in 1971 to 68,416 head in 1991 in the north (Compagnie Ivoirienne du Développement des Fibres Textiles [CIDT], 1993, p. 23). Albert Kientz (1991, p. 23) estimated the annual growth rate of the combined sedentary and transhumant cattle population to be 6.51% between 1968 and 1989. Fulβe transhumant cattle grew at a faster rate (9.1%) than peasant-owned sedentary cattle (5.21%).

The earliest livestock census data for Katiali date from 1982 when there were 3,310 Fulβe cattle in the area. By 1992, their numbers more than doubled to 7,033 cattle. Over the same period, peasant-owned cattle rose from 578 to 1,417 head. Oxen ownership also soared, rising from four to 381 head between 1975 and 1992. By the early 1990s, there were close to 9,000 head of cattle grazing in the Katiali region. Eighty percent of the cattle were Fulβe-owned. The important point for this chapter is that grazing pressure has significantly increased in the region as a result of expanding cattle numbers. Some of the consequences of this pressure on savanna vegetation are examined in the following sections. First we compare the views of environmental change presented in the Côte d'Ivoire NEAP with those elicited from peasant farmers and herders in a survey focused on land use and land cover change.

NARRATIVES OF ENVIRONMENTAL CHANGE

The NEAP Report

According to the World Bank, the NEAP process involves four stages: the identification of environmental problems and their underlying causes; setting priorities; establishing goals and objectives; proposing new policies, institutional and legal reforms, and priority actions. The World Bank considers this process to be straightforward. "It is relatively easy to identify problems and formulate appropriate responses to them" (Greve et al., 1995, pp. 8–16). The more difficult phase, it argues, centers around the implementation of reforms and other policy actions.

What image of environmental change in the savanna emerged from this NEAP process? The NEAP report for the Korhogo region presents a grim scene of environmental degradation in which peasants and pastoralists are blamed for the deforestation of wooded savannas. According to its authors, "Vegetative cover is declining due to the practice of shifting cultivation, bush fires, and the anarchic exploitation of forests and overgrazing" (République de la Côte d'Ivoire [RCI], 1994, p. 14). This change in plant cover is specifically characterized by "a replacement of the tree savanna by the grass savanna." The report goes on to paint a portentous picture of this process: "More generally, the progressive widening towards the south of [grassy] plant formations has climatic repercussions [temperature, rainfall . . .] which in turn affects vegetation cycles" (RCI, 1994, p. 14). This image of a southerly advancing boundary of a vegetation type associated with more arid climates is similar to the "marching desert" view found in the desertification literature. That is, in the absence of field studies, it is assumed that the forms of environmental transformation purportedly taking place in the Sahel are also occurring in humid savannas. Bush fires and indigenous land rights systems are singled out in the report as major forces in this narrative of environmental transformation which Aubréville referred to as "savannization" (Aubréville, 1938).

> We are increasingly witnessing a decline in vegetation cover, essentially due to bush fires used in an intensive and excessive manner leading to the disappearance of certain varieties. . . . These assaults on the environment are essentially linked to the abuses of customary rights, the exaggerated interpretation of the declarations of (the former) President Félix Houphouet-Boigny such as: "*Land belongs to the person who improves it*" and "*that which has been planted by the hand of man must not be destroyed, no matter where*"; and especially the absence of a rural land code. (RCI, 1994, p. 12)

According to this analysis, and that of the World Bank (Cleaver & Schreiber, 1994, pp. 8–10), "traditional" land rights systems are inadequate for the task of modern environmental planning. They may have worked in the past but contemporary land conflicts and insecurity have supposedly prevented farmers from investing in land improvements that might increase agricultural output and improve conservation practices. The assumption is that only when customary rights become commodified in the form of land titles will the incentive to conserve natural resources exist. Indeed, the transformation of land-holding systems to freehold tenure is considered in the Ivorian NEAP to be an important step towards addressing all sorts of environmental problems. Figure 6.2 illustrates this modernization model of tenure change, environmental conservation, and agricultural growth that informs the Ivorian NEAP and which is at the heart of World Bank rural development policies in Africa. These policies only have meaning with reference to agricultural stagnation and/or environmental decline depicted on the vertical axes. The desertification and savannization narratives serve such a need.

How accurate is this image of environmental change in the Ivorian savanna as depicted in NEAP documents? Has the expansion of livestock raising and the area under cultivation led to an increase in grass savannas and a decline in tree cover? Is fire the great destructive force that environmental planners believe it to be?

Research undertaken in subhumid savannas of Cameroon (Boutrais, 1995, Vol. 1, pp. 355–376) and Ethiopia (Coppock, 1993) indicates that heavy grazing pressure and burning commonly result in savannas becoming *more*, not *less*, wooded. That is, in contrast to arid and semi-arid areas where heavy grazing pressure leads to a greater exposure of soils to erosive forces, humid savannas show an inverse tendency toward greater vegetative cover. Woody encroachment has the beneficial effects of protecting soils from erosion and improving soil structure and fertility (Coppock, 1993, p. 55). Do these distinctive

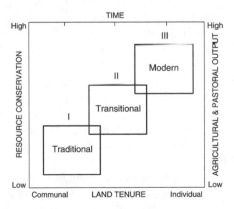

FIGURE 6.2. The modernization model of tenure change, environmental conservation, and agricultural growth.

attributes of humid savannas provide the basis of constructing an alternative environmental change narrative? If so, what might its implications be for environmental policy? The example of Katiali provides some answers to these questions.

Land User Counternarratives

Local perceptions of environmental change were elicited from farmers and herders through a combination of focus group discussions, structured and unstructured interviews, walks through the bush, and interviews with selected elders on the environmental history of the region. We first interviewed a sample of 38 farming households with whom one of us (TJB) has worked for than more 15 years. Five questions were systematically posed to this sample.

1. When you go out into the bush today, is it the same or different in comparison to when you were young?
2. If it is different, what is the difference and how do you explain this change?
3. There are more cattle in the bush today than there were 20 years ago. What has been the impact of these animals on the bush?
4. Are bush fires the same today as they were when you were younger, or are they different?
5. Is it more or less easy to find firewood today in contrast to when you were young, or has it remained the same?

We were surprised and intrigued by the responses given to the first question. A total of 68% of the respondents stated that the bush had become more wooded. Our informants repeatedly told us stories about how the landscape used to be more open in the past, that one could see for great distances, but now this was impossible because of so many trees.

When asked to explain the origins of this more wooded landscape, more than two-thirds of the sample linked it to the arrival of Fulβe herds in the region. Respondents associated the increasing number of cattle to a decline in grass cover, which has allowed trees to expand in numbers. The mechanisms most often described to explain the extension of trees were the dispersal of tree seeds in cattle manure and less intense bush fires. About a quarter (21%) of the sample linked changes in the proportion of woody to herbaceous species to less aggressive bush fires.

Responses to the third question, regarding the impact of cattle on the bush, echoed those to the second. More than half (56%) of the respondents declared that there are more trees because there are more cattle. Some (10%) associated Fulβe cattle with the appearance of new weeds in their fields like *Commelina benghalensis* and *Euphorbia heterophlla*. Another quarter noted that there is less grass with which to thatch houses due to grazing pressure and grass fires set by pastoralists. The principal grasses used for roof construction, *Andropogon gayanus*, *Andropogon schirensis*, and *Schizachyrium sanguineum*, are highly palatable to cattle (Hoffman, 1985, p. 166). These tall, lignified grasses are burnt during the early dry season when animal nutrition is at a stress point (Boudet, 1991, pp. 156–159).

The greatest consensus came in response to the question focused on the nature of bush fires. Ninety-two percent of households agreed that bush fires had changed. They were reported to be occurring earlier in the dry season, to be more frequent, and to be less violent. Most respondents attributed the less aggressive fires to Fulβe cattle, saying that their animals "have eaten up all the grasses." Others noted that the presence of more trees and the expansion in cropping area further reduced the fuel that would normally produce extremely hot fires. In the past, fires were set later in the dry season (February–March) by hunters to enable them to see game more easily. Today, pastoralists set fires as early as October as a range management tool to obtain pasture regrowth.[3] Since grasses are not thoroughly desiccated at this time, they do not burn as hot as they would in the late dry season.

Responses to the fifth and final question on fuel wood availability seemed to contradict the above views. Almost two-thirds of the sample of 23 women declared that it had become more difficult to find wood due to population growth and the expansion of area under cultivation. Ten percent of the respondents did not see any changes in the general availability of fuel wood.

To summarize the results of the household survey, a majority of farmers believe that the most striking change in the savanna environment is that it has become more wooded over the past 30 years. They attribute this increase in trees and shrubs to the growing number of livestock, especially Fulβe cattle, and less aggressive bush fires. Grazing pressure and early dry season fires combine to reduce the intensity of burns. The shift in the proportion of woody and herbaceous plants suggests a major change in savanna ecology due to a changing fire regime.

In three separate group interviews, Fulβe herders commented on the expansion of both fields and trees in the landscape. S. Sangaré (personal communication, 13 August 1995, Katiali) exclaimed that "before you could go all the way to Tiébila [15 kilometers to the southeast of Katiali] without seeing a field. Now all that has changed. There are more fields and more cattle. . . . The cattle have eaten the grass which has allowed trees to grow." Informants declared that there are too many cattle in the region. A major factor behind the influx of cattle is the pastoral dam constructed in 1983 1 kilometer southwest of Katiali which provides a year-round supply of good drinking water. Another factor is the decline in tsetse flies in the region (Erdelen, Müller, Nagel, Peveling, & Weyrich, 1994). In summary, like farmers, herders believe that the savanna has become more wooded in recent decades. They attribute the decline in good forage grasses to the combined expansion of cattle and cropland in the region.

FIRE, CATTLE, AND SAVANNA ECOLOGY

Savannas are plant communities characterized by a continuous grass layer in which trees and bushes of varying height and density are found (Menaut, 1983). In natural savannas (i.e., regularly burned and lightly grazed lands), perennial grasses account for up to 95% of the grass cover. The height and extent of ligneous cover is the criteria most often used in the nomenclature to classify savanna plant communities (Table 6.1). In northern Côte d'Ivoire, all of the vegetation types listed in Table 6.1 are present and usually distributed in mosaic form across the landscape. Figure 6.3 illustrates a typical soil catena of the Katiali region and its associated plant communities. This cross section does not capture the common mosaic pattern of savanna vegetation. The distribution of plant communities is affected by a number of human and biophysical influences that include shifting cultivation, grazing, and fire, as well as topography, soils, and rainfall. Fire is considered to be so important in the maintenance of savannas that they have been called "pyroclimatic" formations. It is widely believed that in the absence of fire, savannas would become increasingly wooded. Controlled burning in experiment stations in Côte d'Ivoire and Nigeria show that parcels protected from fire evolve toward forests, while those subjected to extremely hot, late dry season fires support very few trees. When fires are set during the early part of the dry season, fire-sensitive trees have a better chance of surviving a burn and tend to compete more aggressively with grasses (César, 1994; Gillon, 1983, pp. 626–627) (Figure 6.4). During the early part of the dry season, grasses are not completely desiccated and will not burn

TABLE 6.1. Savanna Vegetation Types

Name (English/French)	Tree/shrub height (meters)	Percent of canopy cover
Open forest/*forêt claire*	8–20	60–80%
Savanna woodland/*savane boisé*	5–18	40–70%
Tree savanna/*savane arboré*	8–15	20–50%
Dense shrub savanna/*savane arbustive dense*	2–8	20–50%
Open shrub savanna/*savane arbustive claire*	2–8	15–20%

Sources: Riou (1995, pp. 45–46, 56–65, 96–100); Mitja (1992: 64–72).

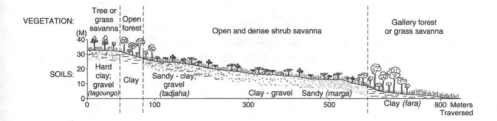

FIGURE 6.3. Vegetation cover in relationship to soil and slope characteristics in the Katiali region, Côte d'Ivoire.

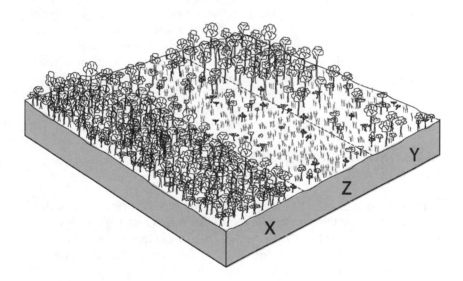

FIGURE 6.4. The impact on vegetation of fire control experiments at Konkondékro (Côte d'Ivoire). When fire is prohibited (parcel X), the savanna evolves toward a dense forest; intense late-dry season fires in parcel Z results in either a grass or open shrub savanna; less violent early-dry-season fires lead to more wooded savannas in parcel Y. *Sources*: Adjanohoun (1964, pp. 131–132); Monnier (1974, pp. 68–69).

as hot as they would if burned in the latter part of the dry period (Gillon, 1983, pp. 618–619).

In addition to the time of year in which burning takes place, fire frequency and grazing pressure also influence fire intensity and rates of bush encroachment. Heavy grazing reduces the amount of combustible material that will burn. Shrubs compete more effectively with grasses for soil nutrients and water when grasses are weakened by repeated grazing (Gillon, 1983, pp. 636–637). Infrequent fires (e.g., once every 3–5 years) will produce higher fuel loads and more intense burns. More frequent (annual) burning will have the opposite effect. In short, savanna ecologists have found that frequent and early dry season fires combined with heavy grazing are conducive to tree and shrub encroachment.

To summarize: farmer and herder perceptions of environmental change are supported by extensive research on savanna ecology and range management in the tropics. Both the general literature and that specifically focused on northern Côte d'Ivoire demonstrate clear links between increased grazing pressure and changing fire regimes on the competition between woody and herbaceous species. Local perceptions and the scientific literature agree that woody species become dominant in landscapes where heavy grazing and less intense fires have allowed tree and shrub seedlings to establish themselves. In the following section we present the results of aerial photo interpretation to determine whether one can detect a shift in the proportion of woody versus grassy savanna types between the 1950s and the late 1980s.

AERIAL PHOTO INTERPRETATION
OF VEGETATION CHANGE

Aerial photographs of Katiali were available for 1956 and 1989 at the scale of 1:50,000. However, the coverage for the 2 years is different. Only a 65-square-kilometer portion of the Katiali territory was included in the 1989 photographs. Moreover, the relatively small scale of these photos hinders interpretation in two significant ways. First, it is impossible to distinguish between wooded savannas and open deciduous forests. Similarly, one cannot differentiate tree savannas from dense shrub savannas at a scale of 1:50,000. As a result, we had to collapse these four savanna plant formations into two categories: *woodlands* for wooded savannas and open deciduous forests and *tree–shrub* for tree savannas and dense shrub savannas.

Second, the extent of bush encroachment is not evident in these photos because of the difficulty in distinguishing the height of woody plants. Thus, what might appear on photos as woodlands may in fact be a tree savanna experiencing heavy shrub invasion. On the ground, there is a noticeable difference between a wooded savanna characterized by uniformly tall trees and a bush-encroached tree savanna marked by trees and shrubs of varying height but exhibiting a density comparable to a wooded savanna. These interpretive difficulties point to the limits of aerial photo interpretation as a method of analysis. However, when used in conjunction with the other methods employed in this study (household surveys, focus group discussions, ground truthing, vegetation transects, review of the savanna ecology literature), aerial photos can be a useful interpretive tool.

Land Cover Changes in Katiali, 1956 and 1989

The most striking change in Katiali's land cover is the dramatic drop in open bush savanna and the noticeable increase in area categorized as woodlands (wooded savanna and open deciduous forest) (Figure 6.5). Open bush declined from 46% to 12% between 1956 and 1989, while the area in woodlands increased from 4% to 31%. These findings confirm the perceptions of local residents that the landscape has become more wooded. It has also become more cultivated, as shown by the increase in area in cropland. This category includes both fallow lands for which the field limits were still visible in the photos and land under cultivation. The area in grasslands and gallery forests remained relatively stable, while the area in dense bush declined by 6% .

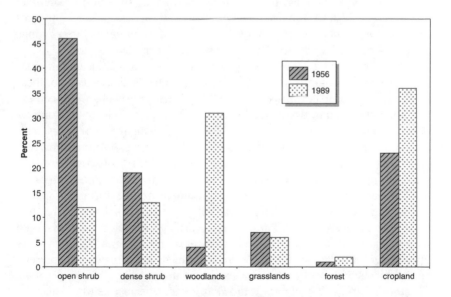

FIGURE 6.5. Land cover changes, Katiali, 1956 and 1989.

To examine more closely the dynamics of vegetation change, we visited a number of locations in the Katiali area where informants indicated the landscape had become more wooded.[4] We were shown two types of wooded places. The first was located within a 3-kilometer radius of the village where both agricultural and grazing pressure is relatively high. Vegetation transects 50 meters in length were run along different sections of the soil catena in the area locally known as "Kafongon." Figure 6.6a shows the results of one of these transects located 1 kilometer from the village in a 15-year-old fallow field situated at the upper slope on sandy–gravely soils. Before being cultivated in the late 1970s and early 1980s, it was an open tree savanna dominated by the economically useful and protected *Parkia biglobosa* and *Vitellaria paradoxa* and perennial savanna grasses such as *Schyzachirium sanguineum*. One informant said that when he was a child in the late 1950s and early 1960s, he could easily see into the distance because there were so few trees. He also mentioned that fires were so intense they would light up the night sky (personal communication, A. Koné, 30 June 1998, Katiali). In 1998 the site was obviously closed in on all sides and experiencing shrub invasion. It was described by Jean César, an authority on range management of tropical savannas who visited the Kafongon site with us, as a "bush-invaded wooded savanna that is becoming a dense open forest dominated by *Isoberlinia doka* but with an understory [i.e., grass layer] that does not correspond to its age" (personal communication, 30 June 1998, Katiali). According to César, the site should have been dominated by *Andropogon gayanus*. However, there were only a few tufts of this perennial grass present. Instead, the site was dominated by the less palatable annuals *Tephrosia pedicellata*, *Hyptis suaveolensis*, and *Cassia mimosidies*. From a range manager's perspective, the herbaceous layer was "extremely degraded."

The second type of wooded place we were shown was located in areas more distant (> 3 kilometers) from Katiali frequented by Fulβe herds. In these locations (e.g., Loupka transect; Figure 6.6b), trees and shrubs were seen expanding out from the edge of small clumps of trees to form larger agglomerations. Even in the absence of grazing pressure and changing fire regimes, trees expand in savannas in this manner (Skarpe, 1991). Informants noted, however, that the agglomeration process had accelerated and that trees had succeeded in establishing themselves beyond the edge of these groves out into open areas. This became apparent to us when we examined the tree and grass species dominating the Loupka site—a 20 year fallow with sandy–clay soils that had been cultivated between 1973 and 1979. According to César, the fallow should have been dominated by *Andropogon gayanus* and savanna perennials like *Hyparhennia dissoluta*. Although *Andropogon* and *Schyzachirium* were present, annual plants were everywhere, especially the unpalatable *Ctenium newtonii*, *Elionuros ciliaris*, and *Tephrosia pedicellata*. After 20 years, one would also expect to find trees like *Isoberlinia* 15–20 meters in height. Instead, all of the trees

a

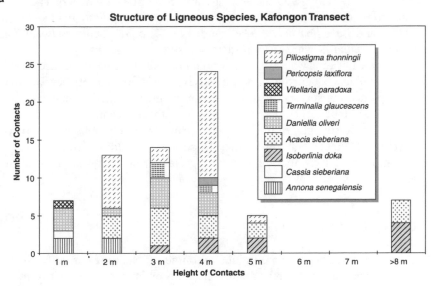

b

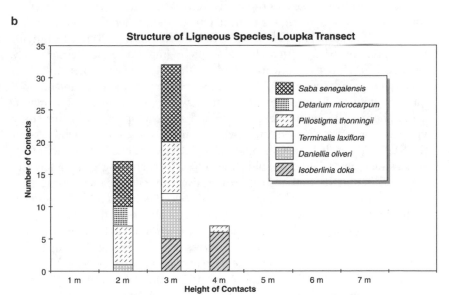

FIGURE 6.6. (a). Structure of ligneous species in 1997, Kafongon transect, Katiali. (b) Structure of ligneous species in 1997, Loupka transect, Katiali.

we encountered along a 50-meter-long transect were below 4 meters. The three most common species were *Dicrostachys cinerea* (34%), *Piliostigma thonningii* (27%), and *Isoberlina doka* (20%). Audru (1977, p. 31) identifies *Isoberlinia* and *Piliostigma* as aggressive invaders of disturbed savannas.

These invading woody species would normally have been checked by late dry season fires but grazing pressure and changes in fire intensity have allowed these pioneering species to survive. Perennial grasses, with their dense root systems and tall height, compete vigorously with tree seedlings for water, soil nutrients, and sunlight. As grazing pressure increases, the root systems do not develop and grass height diminishes. Under these conditions, trees can compete more effectively and will spread, especially if bush fires are not too intense.

CONCLUSION

Given the extraordinary amount of environmental planning currently underway in Africa and its far-reaching implications on land use, access, and management, one obvious conclusion of this study is that further research on environmental change dynamics is critical. One of the challenges in confronting the environmental data problem is that so little data exist with any meaningful time depth. Even where sources like aerial photographs do exist, their relatively small scale rarely permits one to make more than very general statements. This situation demands that multiple approaches be pursued to determine the spatial and temporal dynamics of environmental change that are not apparent in aerial photos. The multiscale, multimethod approach pursued in this study yielded different results from the so-called participatory approach followed in the NEAP process in which the opinions of selected individuals from different social strata were solicited in public meetings. Not surprising, peasant farmers and herders were reticent in such forums. Even if they were more vocal, their interpretations would not be taken seriously because they would not conform to the discursive frameworks and simplifying narratives of donor agencies.

By contrasting local understandings of environmental change with the Côte d'Ivoire NEAP desertification narrative, we raise a fundamental question: Why does the NEAP view dominate despite contradictory evidence? In their discussion of regional discursive formations, Peet and Watts (1996, pp. 15–16) leave open the questions of how and why certain themes and images appear and disappear, "only to reappear with even greater intensity in new guises." Following Goldman (2001), we argue that the scientific "truth" of desertification is given legitimacy by the networks of power in which the idea circulates. The NEAP process recycles certain ideas of environmental change because the aid donor community has already indicated that they are important. Developing country governments dependent on aid monies must demonstrate that they are considering these important environmental issues even if they are imaginary ones. Alternative ideas of environmental change are not taken seriously because they do not have the requisite institutional authority and thus the ability to attract funds that will empower government ministries. The terms of reference and tidy templates of the NEAP process guarantee that the knowledge produced will conform to the World Bank's parameters and realms of intervention. The desertification narrative circulates within these networks of power. It appears in the World Bank's environmental strategy paper for Côte d'Ivoire (World Bank, 1994), in NEAP planning documents (RCI, 1994), in the speeches of government ministers (*Fraternité Matin*, 1999), and in the news magazines of NGOs (Gomé Gnohite, 1998) whose funding depends on their participation in the "development" process. In sum, the global narrative of desertification remains the dominant explanation of environmental change in the Ivorian savanna because it empowers certain people and institutions who have a stake in it remaining "everyone's business."

ACKNOWLEDGMENTS

The field research on which this chapter is based was supported by the John D. and Catherine T. MacArthur Foundation. In addition to the coauthors, the collaborative research team included Mameri Camara, Sinali Coulibaly, and Tiona Ouattara. We thank our Ivorian collaborators as well as the following individuals for their comments on earlier drafts of this chapter: David Anderson, Jean César, Zoumana Couilbaly, Leslie Gray, Alex Winter-Nelson, and Karl Zimmerer.

NOTES

1. The IDA is the "soft loan window" of the World Bank Group that offers loans to member countries with average annual per capita incomes of $1,465 or less at concessionary rates.
2. Measurements were made every 50 centimeters along the 50-meter transect. If a woody plant came into contact with the 8-meter pole at this interval, we noted the species of plant and the height at which it touched the pole. Figures 6.6a and 6.6b shows the number of times a species touched the pole (y-axis) at a specific height (x-axis).
3. Fire is also used to control vectors of animal diseases such as ticks and tse tse flies (Gillon, 1983, pp. 630, 636).
4. Our Senufo and Jula informants distinguished between two general types of savanna landscapes: wooded and open. The Senufo say *tétungo* to indicate "a place where there are a lot of trees," in contrast to *téfigue* or "open place." In Jula, the comparable terms are *a yoro tura* and *a yoro kenegbe*.

REFERENCES

Adams, W. M. (1995). Green developent theory. In J. Crush (Ed.), *Power of development* (pp. 87–99). London: Routledge.

Adjanohoun, E. (1964). *Végétation des savanes et des rochers découvertes en Côte d'Ivoire Centrale* (Mémoire ORSTOM 7). Paris: Editions ORSTOM.

Anderson, D., & Grove, R. (1987). Introduction: The scramble for Eden: Past, present and future in African conservation. In D. Anderson & R. Grove (Eds.) *Conservation in Africa: People, policies and practice* (pp. 1–12). Cambridge, UK: Cambridge University Press.

Aubréville, A. (1938). *La forêt coloniale: Les forêts de l'Afrique Occidentale Française* (Annales, Tome 9). Paris: Académie des Sciences Coloniales.

Audru, J. (1977). *Les ligneux et subligneux des parcours naturels soudano-guinéens en Côte d'Ivoire* (Note de Synthèse No. 8). Maisons-Alfort, France: IEMVT.

Bassett, T. J. (1986). Fulani herd movements. *Geographical Review, 76*(3), 233–248.

Bassett, T. J. (1993). Land use conflicts in pastoral development in northern Côte d'Ivoire. In T. J. Bassett & D. Crummey (Eds.), *Land in African agrarian systems* (pp. 131–154). Madison: University of Wisconsin Press.

Batterbury, S., Forsyth, T., & Thomson, K. (1997). Environmental transformations in developing countries: Hybrid research and democratic policy. *Geographical Journal, 163*(2), 126–132.

Blaikie, P. (1994). *Political ecology in the 1990s: An evolving view of nature and society* (CASID Distinguished Speaker Series, No. 13). East Lansing: Michigan State University.

Blaikie, P. (1995). Changing environments or changing views? *Geography, 80*(348), 203–214.

Boudet, G. (1991). *Manuel sur les pâturages tropicaux et les cultures fourragères* (IEMVT Collection Manuels et Précis d'Elevage 4). Paris: La Documentation Française.

Boutrais, J. (1995). *Hautes terres d'élevage au Cameroun* (2 vols.). Paris: Editions ORSTOM.

César, J. (1994). Gestion et aménagement de l'espace pastoral. In C. Blanc-Pamard & J. Boutrais (Eds.), *A la croisée des parcours* (pp. 111–145). Paris: Editions ORSTOM.

César, J., & Zoumana, C. (1995). *Comparaison de troupeaux mono et pluri-spécifiques sur une végétation de savane soudanienne à Korhogo (Côte d'Ivoire)* (Compte Rendu Technique No. 2, Natural Resource Development and Utilization in the Sahel [IDESSA/CIRAD-EMVT]). Bouaké, Côte d'Ivoire: IDESSA.

Cleaver, K., & Schreiber, G. (1994). *Reversing the spiral: The population, agriculture, and environment nexus in sub-Saharan Africa*. Washington, DC: World Bank.

Compagnie Ivoirienne du Développement des Fibres Textiles. (1993). *Rapport annuel des activitiés*. Bouaké, Côte d'Ivoire: Author.

Coppock, D. L. (1993). Vegetation and pastoral dynamics in the southern Ethiopian rangelands: Implications for theory and management. In R. Behnke, I. Scoones, & C. Kerven (Eds.), *Range ecology at disequilibrium* (pp. 42–61). London: Overseas Development Institute.

Crush, J. (Ed.). (1995). *Power of development*. London: Routledge.

Daget, P., & Poissonet, J. (1971). Une méthode d'analyse phytologique des prairies: Critères d'application. *Annales Agronomiques, 22*(1), 5–41.

Erdelen, W., Müller, P., Nagel, P., Peveling, R., & Weyrich, J. (1994. *Implications écologiques de la lutte anit-tse-tse en Côte d'Ivoire nord et centre* (GTZ Project No. 87.2539.2. Rapport Final). Sarrebruck, Germany: GTZ.

Fairhead, J., & Leach, M. (1998). *Reframing deforestation: Global analyses and local realities: Studies in West Africa*. London: Routledge.

Falloux, F., & Talbot, L. (1993). *Crisis and Opportunity: environment and development in Africa*. London: Earthscan.

Forsyth, T. (2001). Critical realism and political ecology. In J. Lopez & G. Potter (Eds.), *After Postmodernism: An introduction to critical realism* (pp. 146–154). London: Athlone.

Fraternité Matin. (1999, July 5). Reconstituer le patrimoine forestier.

Ferguson, J. (1990). *The anti-politics machine: "Development," depoliticization, and bureaucratic power in Lesotho*. Minneapolis: University of Minnesota.

Gillon, D. (1983). The fire problem in tropical savannas. In F. Bourlière (Ed.), *Ecosystems of the world: Vol. 13. Tropical savannas* (pp. 617–641). Amsterdam: Elsevier.

Gomé Gnohite, H. (1998). Ce que je pense. . . . *Bulletin d'information de la sensibilisation envrionnementale, 2*, 14.

Goldman, M. (2001). The birth of a discipline: Producing authoritative green knowledge, World Bank-style. *Ethnography, 2*(2), 191–218.

Greve, A. M., Lampietti, J., & Falloux, F. (1995). *National environmental action plans in Sub-Saharan Africa* (Towards Environmentally Sustainable Development in sub-Saharan Africa, Paper. No. 6). Washington, DC: World Bank.

Hoffman, O. (1985. *Pratiques pastorales et dynamique du couvert végétale en pays Lobi* (Nord Est de la Côte d'Ivoire) (Collection Travaux et Documents 189). Paris: Editons ORSTOM.

Kientz, A. (1991). *Développement agro-pastoral et lutte anti-tsé-tsé, Côte d'Ivoire*. Strasbourg, France: GTZ.

Leach, M., & Mearns, R. (Eds.). (1996). *The lie of the land: Challenging received wisdom on the African environment*. Oxford, UK: James Currey.

Menaut, J.-C. (1983). The vegetation of African savannas. In F. Bourlière (Ed.), *Ecosystems of the world: Vol. 13. Tropical savannas* (pp. 109–149). Amsterdam: Elsevier.

Mitchell, T. (1995). The object of development: America's Egypt. In J. Crush (Ed.), *The power of development* (pp. 129–157). New York: Routledge.

Mitja, D. (1992). *Influence de la culture itinérante sur la végétation d'une savane humide de Côte d'Ivoire* (Collection Etudes et Thèses). Paris: Editons ORSTOM.

Monnier, Y. (1974). *Decouverte aerienne de la Côte d'Ivoire*. Abidjan: Editions Photivoire.

Mortimore, M. (1989). *Adapting to drought: Farmers, famines, and desertification in West Africa*. Cambridge, UK: Cambridge University Press.

Peet, R., & Watts, M. (Eds.). (1996). *Liberation ecologies: Environment, development, social movements*. London: Routledge.

République de la Côte d'Ivoire. (1994). *Ministère de l'Environnement et du Tourisme, Plan National d'Action Pour l'Environnement. Cellule de Coordination Synthèse régionale: Région nord (Korhogo)*. Abidjan: PNAE-CI.

Riou, G. (1995). *Savanes, l'herbe, l'arbre et l'homme en terres tropicales*. Paris: Masson/
     Armand Colin.
Scott, J. (1998). *Seeing like a state*. New Haven, CT: Yale University Press.
Skarpe, C. (1991). Impact of grazing in savanna ecoystems. *Ambio, 20*(8), 351–356.
Société d'Etudes pour le Développement Economique et Sociale. (1965). *Région de
     Korhogo: Etude de développement socio-économique: Vol. 3. Rapport agricole*.
     Paris: Author.
Stocking, M. (1987). Measuring land degradation. In P. Blaikie & H. Brookfield (Eds.),
     Land degradation and society (pp. 49–63). London: Metheun.
Stocking, M. (1996). Soil erosion: Breaking new ground. In M. Leach & R. Mearns
     (Eds.), *The lie of the land: Challenging received wisdom on the African environ-
     ment* (pp. 140–154). Oxford, UK: James Currey.
Swift, J. (1996). Desertification: Narratives, winners, and losers. In M. Leach & R.
     Mearns (Eds.), *The lie of the land: Challenging received wisdom on the African en-
     vironment* (pp. 73–90). Oxford, UK: James Currey.
Swyngedouw, E. (1997). Neither local nor global: "Glocalization" and the politics of
     scale. In K. R. Cox (Ed.), *Spaces of globalization: Reasserting the power of the local*
     (pp. 137–166). New York: Guilford Press.
Watts, M. (1987). Drought, environment, and food security: Some reflections on peas-
     ants, pastoralists, and commoditization in dryland W. Africa. In M. Glantz (Ed.),
     Drought and hunger in Africa: Denying famine a future (pp. 149–162). Cambridge,
     UK: Cambridge University Press.
World Bank. (1994). *Côte d'Ivoire: Vers un développement durable* (Rapport No. 13821-
     IVC). Abidjan: Author.

CHAPTER 7

Environmental Zonation
and Mountain Agriculture
in Peru and Bolivia

Socioenvironmental Dynamics of Overlapping
Patchworks and Agrobiodiversity Conservation

Karl S. Zimmerer

A model of tiered zones is often used to guide the environmental management of mountain agriculture and land use worldwide (Buzhuo, Lijie, Haosheng, & Higgitt, 1997; Kappelle & Juárez, 1995). This model assumes that the zonation of resource use in mountain agriculture is distributed into stack-like configurations of bands or belts. The tiered zonation model is based on a common interpretation of the spatial and environmental organization of mountain farming. It presupposes the determining influence of decision making among farmers that optimizes productivity and risk aversion based on elevation-controlled factors of the mountain environment. According to the tiered zonation model, elevation-based zones serve as integrative categories that group the agricultural-yield-determining factors of topography, climate, and vegetation. Widespread assumptions of tiered zonation lead to the familiar representation of the stacked, layer-cake-like units of mountain land use (Figure 7.1).

Adapted by Karl S. Zimmerer from the article "The Overlapping Patchworks of Mountain Agriculture in Peru and Bolivia: Toward a Regional–Global Landscape Model," published in *Human Ecology, 27*(1), 135–165 (1999). Adapted by permission from Kluwer Academic/Plenum Publishers. The adapting author is solely responsible for all changes in substance, context, and emphasis.

137

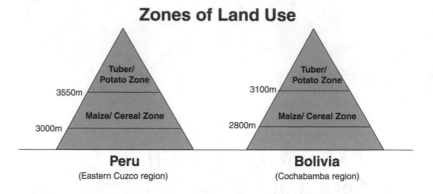

FIGURE 7.1. Standard-style representation of the zonal or "verticality" model of mountain farming and land use. This model assumes discrete zones that are layered according to elevation. The example refers specifically to the predictions of this model applied to the major types of farming in Cuzco, Peru, and Cochabamba, Bolivia.

The tiered zonation model is increasingly utilized in policymaking as well as in discussions and debates that are associated with conservation management in areas of mountain agriculture such as the central Andes of Peru, Bolivia, and Ecuador (Sarmiento, 2000). At the same time, the rethinking of tiered zonation is now being undertaken in conjunction with environmental planning and proposed conservation management (Zimmerer, 2000a, 2002a). Distinguishing the scales at which the tiered model of environmental zonation can be realistically applied to mountain agriculture is one of several scale issues worthy of close examination and possible rethinking (Lauer, 1993; Mayer, 1985, 2002; Troll, 1968). Focusing on the patterns and process of the socio-environmental scaling of land use, this chapter offers a political–ecological analysis of the zonation of mountain agriculture.

Three concepts of political ecology are foundations of this study. First, the concepts and analysis of ecological science are used in order to determine how the dynamics of agricultural land use are related to political and economic factors in farm-level decision making (Blaikie, 1994; see also Bassett & Koli, Chapter 6, and Turner, Chapter 8, this volume). The political ecology of food plant cropping in mountainous farm landscapes is thus seen as involving issues of ecological biogeography in interaction with cultivation practices. This emphasis furnishes the first foundation for examining the spatial–environmental organization of mountain agriculture in the Andean communities of Peru and Bolivia. Closely related to this first foundation are the rationales of decision making, including broadly political economic factors, about land use in mountain agriculture (Painter, 1995). This discussion is focused on the decision making of individual farm households, and devotes special attention to the increasingly diversified style of cultivation that is evermore characteristic of rural and periurban Andean locales. Second, the examination of decision-making rationales is extended to the community and interhousehold level, where land users may coordinate such activities as field cropping or livestock grazing (Blaikie & Brookfield, 1987; Zerner, 1994, 2000). Third, my conceptual focus is the cultural landscape basis of the study. It incorporates ethnography and social theory guided ideas of how residents themselves define spatial–environmental categories through cultural images of landscapes that are produced in their land use, namely, their "Models for Agriculture" (Lefebvre, 1974/1991; Mayer, 2002; see also Robbins, Chapter 9, this volume).

Broad interests of political ecology in the issue of scale, and zonation-based conservation planning and management in particular, also serve as a guide for the design of this study. Zonation-based units of particular types of land or resource use have been a territorial linchpin of the expanding jurisdiction of designated conservation areas in the "conservation boom" of the recent past (Zimmerer, 2000a, 2003c; Zimmerer & Carter, 2002). The tier-based model of zonation in mountain agriculture is of special relevance to this expanding conservation management since it has typically counted as one of the principal examples of zonal land use. During recent decades, understandings of tier-based zonation have filtered into new zonal designations such as "buffer zones," "cultural zones," and "transitional zones," which are bread-and-butter designations of conservation management. This expanded type of scale making is one of the most common means of regulating resource management in parks-with-people, community-based conservation initiatives, integrated conservation and development projects, and biosphere reserves (Stevens, 1997; Wells & Brandon, 1993; Western & Wright, 1994). Tier-based zonation is also evident in the management plans of protected areas that contain large areas of mountain agriculture, including the vast Vilcabamba-Amboró Conservation Corridor that has been newly established in the central Andes of Peru and Bolivia. Such designations as a lower elevation "Maize/Cereal Zone" and an upper elevation "Potato/Tuber Area" are under consideration as spatial backbones of the new conservation management (Figure 7.1).

This study is also designed to address a core set of concepts in cultural–human ecology and agroecology–biogeography that are closely related to current interests in political ecology (Bassett & Zimmerer, 2003). In cultural and human ecology, the idea of environmental zonation and its depiction in mountain agriculture and land use is a major area of study (e.g., Brush, 1976). This idea of human–environmental interaction is premised on the determining influence of the decision making of mountain farmers that optimizes agricultural productivity and risk aversion based on elevation-controlled environmental factors. Central to the policy concerns of cultural and human ecology is the idea of *in situ* agrobiodiversity conservation, which holds that continued small farm cultivation is, and should be, a foundation of the evolutionary and agroecological maintenance of diverse food plants (Brookfield & Padoch, 1994; Brush, 1995; Denevan, 2001). The idea of genotype-by-environment interaction is also a core of this study. Founded in the agronomic and ecological sciences, this concept is applied to the nature and extent of genetic influences on the relations of food plant types to environmental factors such as elevation-related climate and soils conditions.

This study draws on original, field-based case studies of rural, Indian peasant communities in Peru and Bolivia in order to evaluate the spatial–environmental predictions of the tiered zonation model of mountain agriculture. Location of the field studies in the Andean countries requires that this study engage with a rich variety of other studies of existing mountain agriculture in the region (Lauer, 1993; Mayer, 2002). A multitude of cultural-environmental interpretations of mountain landscapes, including self-representations, is well developed in traditional studies of the Andean peoples and their societies. These traditions emphasize their customary and frequently complex use of diverse, elevation-related mountain environments, which in general is referred to as "verticality." Findings of the present study demonstrate (1) substantial overlap of the primary spatial units of local farm landscapes with respect to elevation in conditions of compressed "verticality"; and (2) varied degrees of patchiness or cohesion within the farm spaces. This spatial overlap, in conjunction with patchiness or cohesion, is a major contrast to tier-based zonation. In the Conclusion, these findings are used to suggest a model of "overlapping patchworks."

The case studies are based on the Indian peasant communities of Mollomarca (in Cuzco, Peru) and Pampa Churigua (in Cochabamba, Bolivia). In the Mollomarca community, the official population is approximately 350 inhabitants, according to the 1995 community census. Their farmland ranges from near 3,100 meters to 4,100 meters (Figure 7.2). The population of the Pampa Churigua community is estimated at 600 persons. Their farm territory reaches from 2,800 meters to 3,450 meters (Figure 7.3). Both communities are located within microcenters of agrobiodiversity. These communities and many others in neighboring areas are increasingly subject to management for conservation and environmentally sound development. Major crop plants in these communities include diverse Andean potatoes, maize, barley, wheat, ulluco, quinoa, fava bean, and tarwi.

MODELS OF LANDSCAPES

The first focus of my research was the location of crop types within community landscapes. Field methods included semistructured interviews, ethnographic, participant observation, and agroenvironmental field sampling that were integrated in a multimethod design (Zimmerer, 2002a). During a total of

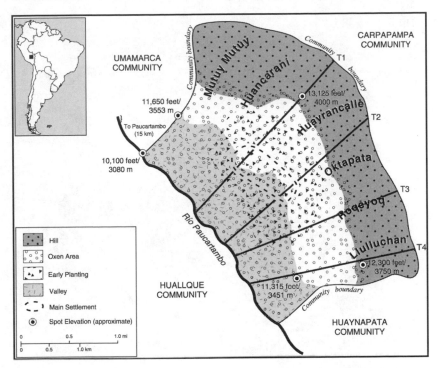

FIGURE 7.2. Farm units of mountainous landscape in the Mollomarca community, Cuzco, Peru. Units are defined in socioenvironmental terms among farmers and in research analysis. Overlap distinguishes the spatial pattern of the four basic units of agriculture: hill, oxen area, early planting, and valley. Patchworks result from the partial cohesion of each farm unit.

20 months in Mollomarca and other Cuzco communities, I asked persons about food plant production and consumption and the location of their fields. The field interviews were conducted in Quechua and Spanish with adult men and women using the help of a field assistant. Detailed interviews were conducted with 15 families of the Mollomarca community whose residences and fields are scattered across community territory. The 15 persons consisted of 12 men and three women.

Persons were also asked which crops were sown previously and which crops would be planted subsequently. I used similar methods in the Pampa Churigua community during 14 months of field study. A pair of minor adjustments followed guidelines for multisite ethnography (Bernard, 1994). In Mollomarca, where farming is the economic mainstay, persons easily recounted crops that were rotated during the previous 3 years. Therefore inter-

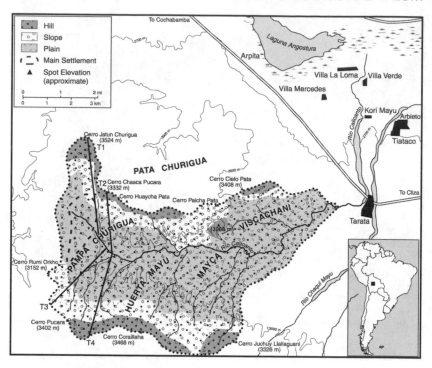

FIGURE 7.3. Farm units of mountainous landscape in the Pampa Churigua community, Cochabamba, Bolivia. Overlap occurs among the three farm units: hill, slope, and plain. (Main settlement shown only for Pampa Churigua community.)

views adopted a 3-year time frame. In Pampa Churigua, by contrast, farmers recalled with confidence the prior 2 years of cropping, a shortened recollection that reflects their typically greater involvement in nonfarm work and migration. Interviews in the latter community thus covered a 2-year time frame. Another regional difference was that 29 Pampa Churigua farmers (19 men, 10 women) were interviewed, since they hold access to fewer fields.

Familiarity with the details of local topography, landmarks, and named places were used to identify each field as belonging to either valley bottom and lower slopes (predicted to be the Maize/Cereal Zone) or upper slopes (predicted to be the Potato/Tuber Zone). Elevation limits of zones were specified using studies of other communities in eastern Andean areas of Peru and Bolivia (e.g., Brush, 1976; Lauer, 1993). In the Mollomarca community, limits of the presumed Maize/Cereal Zone (3,100–3,550 meters) and the Potato/Tuber Zone (3,550–4,000 meters) reflect the elevation span of community territory (Figure 7.1). In the Pampa Churigua community, the limits reflect moderate compression (Maize/Cereal Zone, 2,800–3,100 meters; Potato/Tuber Zone, 3,100–3,450 meters), which is typical of the territories of Indian peasant communities in Cochabamba.

Results show that the 15 families in the Mollomarca community managed a total of 114 field sites, 47 in the predicted Maize/Cereal Zone and 67 in the Potato/Tuber Zone (mean = 7.6 fields/household) (Table 7.1). Their 29 counterparts in the Pampa Churigua community held rights over a total of 94 fields (mean = 3.2 fields/household). Responses about current cropping and crop rotations yielded a total 570 field-year pairs in Mollomarca and 376 field-year pairs in Pampa Churigua. Total data on crop location thus consists of 946 field-year pairs. Of these, 411 field-year pairs (235 in Mollomarca, 176 in Pampa Churigua) belonged to cultivation sites in the bottoms and lower slopes of valleys (Maize/Cereal Zone). Correspondingly, 535 field-year pairs (335 in Mollomarca, 200 in Pampa Churigua) belonged to planting sites on the upper slopes (Potato/Tuber Zone).

Each presumed zone is planted with a mixture of crop types. Of 235 field-year pairs that were sown in the Maize/Cereal Zone of the Mollomarca community, these particular crops were found in 128 fields, or 54.5 % (Table 7.1). In addition, 57 potato fields and 14 parcels (total = 30.2%) planted with other

TABLE 7.1. Crop Zones, Field Environments, and Food Plants of the Mollomarca Community (Cuzco, Peru) and the Pampa Churigua Community (Cochabamba, Bolivia)

Zone (community)	Elevation range (meters)	Total fields	Mean fields/ person	Total field-years	Maize	Potato	Lisas	Barley	Wheat	Other crops	Fallow
Maize/cereal zone											
Mollomarca (Paucartambo)	3,100–3,550	47	3.1	235	94	57	7	28	6	7	36
Pampa Churigua (Cochabamba)	2,800–3,100	44	1.5	176	51	43	0	16	31	14	21
Zone total		91		411	145	100	7	44	37	21	57
Potato/tuber zone											
Mollomarca (Paucartambo)	3,550–4,000	67	4.5	335	0	73	1	34	1	44	182
Pampa Churigua (Cochabamba)	3,100–3,450	50	1.7	200	65	35	0	9	42	3	46
Zone total		117		535	65	108	1	43	43	47	228
Total (both zones)		208		946	210	208	8	87	80	68	285

tubers and legumes (ulluco, fava beans, tarwi) also occurred in the area predicted to be the Maize/Cereal Zone. Of 176 field-year pairs in the Pampa
Churigua community's predicted Maize/Cereal Zone, 98 fields (55.7%) were
occupied by maize and cereals while other crop types were planted in 57 fields
(31.8%). Fallow fields totaled 36 (15.3%) in Mollomarca and 21 (11.9%) in
Pampa Churigua. In general, the proportion of fields sown with other crop
types, that is, not predicted by the tiered zonation model, was greater than 30%
overall and was more than half the total number of fields in the predicted
Maize/Cereal Zone.

Mixtures of crop types also are characteristic of the predicted Potato/Tu-
ber Zone. In the Mollomarca community, 335 fields of this area included 74
fields (22.1%) of potato and tuber crops (Table 7.1). Major nontuber crops
consisted of 35 fields (10.4%). Fallow fields (54.3%) and mixed minor crops
(11.6%; legumes, tubers, pseudocereals) also occurred. In the area of the
Pampa Churigua community that was predicted to be the Potato/Tuber Zone,
which included 200 sampled fields, potatoes and tubers accounted for 38 fields
(19.0%). Fallow fields totaled 46, or 23%. Nontuber crop types made up 116
fields, or 58% of the total. These results show that crop types that are presumed
to be atypical actually amount to 50% or more of plantings. As seen, predictions are overturned completely in the case of Pampa Churigua's predicted Po-
tato/Tuber Zone. (As discussed further below, the community's relatively
moderate span of elevations [2,800–3,450 meters] permits a high degree of
mixing of crop types.)

The second focus was to investigate the location of crop types both within and among presumed zones. Points were sampled at 100-meter elevation in-
tervals along a set of four cross-community transects (Table 7.2, Figures 7.2,
7.3). (Elevation was estimated with a Thommen TX-16 mechanical altimeter/
barometer.) At each point, I noted the crop types of the five nearest fields and
recorded them in sequence, as shown in Table 7.2. Fields estimated to be on
contour were considered. Results demonstrate the elevation-related overlap of
crop types, where overlap is defined as cooccurrence. In the Mollomarca community, overlap was observed in the major crop pairs as follows: barley-maize
(> 500-meter overlap), barley-potatoes (> 600-meter overlap), and maize-
potatoes (> 200-meter overlap). In the Pampa Churigua community, the major
crops (maize, potatoes, wheat, barley) overlapped between 2,800 and 3,200
meters.

Patchiness, defined as contiguous locations, was also observed in the field
sampling (Table 7.2). Patchiness was found both along the elevation transects
and along contours at a single elevation. Along transects, crop type patchiness
occurred in elevation-related ranges. Transects across the landscape of the
Mollomarca community demonstrated clusters of maize fields (3,100–3,500
meters), barley fields (3,100–3,900 meters), and potato fields (3,300–4,000
meters). Along contours, patchiness was found in clusters of crop types that

TABLE 7.2. Field Types along Contours at Transects

Elevation (meters)	Transect 1	Transect 2	Transect 3	Transect 4
Mollomarca				
3,100	f-b-m-m	f-f-m	m-b-m-f-f	m-m-b-f
3,200	f-f-m-m-m	b-fa-fa-t-m	m-m-m-f-f	f-m-b-m-f
3,300	f-m-b-p-p	p-p-m-m-fa	m-m-b-b-f	b-m-m-f-f
3,400	m-m-b-p-p	p-p-b-m-b	m-m-fa-f-f	m-m-f-f-f
3,500	f-f-m-m-p	p-p-p-m-m	b-m-b-m-b	m-b-f-f-f
3,600	f-p-p-p-p	p-p-p-p-p	p-p-b-b-b	b-f-f-f-f
3,700	f-f-f-p-p	p-p-p-p-p	p-p-b-b-b	b-f-f-f-f
3,800	f-f-f-p-p	p-p-p-p-b	b-p-p-f-f	b-b-f-f-f
3,900	f-f-f-p-p	f-p-p-p-b	p-p-p-oc-f	b-l-f-f-f
4,000	f-f-f-f-f	f-f-p-p-p	p-p-oc-l-f	f-f-f-f-f
Pampa Churigua				
2,800	f-m-a-m-a	w-m-p-m-p	f-b-m-w-m	p-m-a-a-p
2,900	m-p-m-m-p	p-m-w-m-w	b-f-m-p-m	f-fa-m-m-w
3,000	f-m-w-m-p	f-w-m-w-w	p-m-w-w-fa	p-m-w-w-fa
3,100	f-m-p-w-m	p-f-m-b-m	w-fa-m-f-m	m-f-m-p-m
3,200	p-p-w-m-p	p-w-p-m-p	m-p-o-f-p	p-m-m-f-f
3,300	p-f-w-f-w	f-f-w-m	w-f-w	p-p-m-m
3,400	w-w-f-b	p-w-w-b	p-p-f-f	p-w-f-w

Note. Symbols: a, alfalfa; b, barley; f, fallow; fa, fava beans; l, lisas (ulluco); m, maize; oc, oca; p, potato; w, wheat. Transect locations are shown on Figures 7.1 and 7.2.

occurred in adjacent or nearby fields (Table 7.2). In Mollomarca, fields of the same crop type clustered moderately along contours. Examples given in Table 7.2 include barley (e.g., 3,300, 3,600, 3,700 meters on T3), potatoes (e.g., 3300–4000 meters on T2), and maize (3,100, 3,200, 3,400 meters on T1). The degree of patchiness is considerably less in the land use of Pampa Churigua community than in Mollomarca. (This difference is due primarily to the extreme diversification of agriculture and nonfarm economic activities in the Cochabamba area of central Bolivia.)

A third focus was to investigate the rationales of farmers for locating their crop types at particular sites within community landscapes. Semistructured interviews with farmers in the Mollomarca and Pampa Churigua communities elicited planting rationales for their currently planted fields (Table 7.3). Since their fallow fields were omitted, the interviews addressed 70 fields in Mollomarca and 78 fields in Pampa Churigua. All responses were recorded, which led the total of rationales (282) to number far greater than the number of fields ($n = 148$).

Common rationales, in order, were good yield (66), market price and profitability (41), custom or tradition (33), maximum yield (33), and desired

TABLE 7.3. Rationales for Locations of Current Sites of Food Plants

	Mollomarca	Pampa Churigua	Total
Custom	25	8	33
Risk	19	10	29
Maximum yield	17	16	33
Good yield	22	44	66
Market price/profitability	14	27	41
Desired foodstuff	16	14	30
Seed availability	9	13	22
Labor availability	7	18	25
Don't know	1	2	3
Total	130	152	282

Note. n = 44 farmers; total of 148 fields.

foodstuff (30) (Table 7.3). Each major rationale about planting location occurred in at least 20% of cases (*n* = 148). "Good yield" is the most common criterion, indicating the general importance of production factors in decision making about field location. The overall commonness of several factors—including "good yield" (as opposed to "maximum yield")—suggests that yield optimization is only moderately influential in decision making. This finding helps to explain why farmers locate their crop types across an agroenvironmental spectrum that is considered acceptable, rather than restrict their plantings to certain optimum sites that would generate a pattern of tier-based zonation.

Rationales about the location of crop types also give evidence of economic links to extraregional processes. In the Mollomarca community, linkages are apparent in the commonness of market price and profitability as a criterion. This factor was applied most widely to the premier commercial crops: market potatoes of the "early planting" and agroindustrial malting barley. Product and input markets for these two crops entail extraregional processes that are influenced by national and international policies discussed in the next section. Complexities of regional–global linkages are evidenced by how several farmers referred to "custom" even for their market and agroindustrial crops, suggesting that reference to tradition can serve to mask innovation. In the community of Pampa Churigua, by contrast, market integration is evidenced most extensively by off-farm migration. The highly varied extent (frequency, duration) of off-farm migration among households contributes to the importance of multiple criteria in decision making about cropping and field location.

Processes leading to the patchiness of crop types in adjacent fields were also studied. Most notable in this regard is the use of crop–livestock coordination known as "common field agriculture" or "sectoral fallow," an arrangement that locally is referred to as *suerte* (Zimmerer, 2002b).

Common field agriculture is a formal management system of crop and livestock coordination that is based on several contiguous areas (typically six–10 sectors in total) which each consist of dozens of individual fields (Mayer, 1985, 2002). These areas, referred to as "sectors," are spatially segregated and their use is regulated and allocated to either cropping or livestock grazing according to community institutions. Use of this technique has led the Mollomarca farmers to choose a single crop such as potato or barley for an entire sector. The six sectors of the Mollomarca community, which each contain more than 50 fields, are labeled in Figure 7.2. Common field agriculture is practiced in more than 20 communities of Indian peasants in the area of eastern Cuzco, where it is adjusted according to extraregional economic pressures and opportunities. (Common field agriculture, or sectoral fallow, has not been utilized in Pampa Churigua and the main Cochabamba region at least during the recent period that begins in 1970.)

Common field agriculture exerts a pronounced effect on the community-level biogeographic patterning of local food plants. This effect infers a mixed influence on zonation-like patterns. On the one hand, the arrangements of common field agriculture, such as in the Mollomarca community, clearly lead to the clustering of fields that contain the same crop. For example, the clustering of potato and barley fields along Transects 2 and 3 at elevations above 3,600–3,800 meters, as shown in Table 7.2, is attributed to the effects of common field agriculture. (The clustering at these elevations of fields that are managed as fallows, most noticeable along Transects 1 and 4, is also attributed to common field agriculture.) On the other hand, the zonation-like pattern of common field agriculture is only partial. Its sectors exhibit limits of elevation that typically are uneven and varied, even at the intracommunity scale. Local differences occur at both the lower reaches (3,600–3,900 meters) and the uppermost locations (4,000–4,150 meters) where fields are cultivated. Thus, while common field agriculture is important as a form of community resource management that clusters land use, it corresponds only partly to the model of zone-style agriculture.

Various immediate rationales are active in the uneven change of the elevation-related limits of common field agriculture. Lifting of its lower elevation limit is often due to the increased economic value of lower elevation field sites, primarily as a result of intensified potato production and agribusiness-based growing of malting barley. The medium-term duration of managed fallow, which is strictly enforced through community institutions and customs, is not considered practicable in certain communities due to within-community dissent. Poorer farm households and young families (in essence, those with smaller farm areas) are often those least able to incorporate this fallow period. These groups have, at times, become effective in pressuring their communities for either the reduction or elimination of common field agriculture. Addressing their access to adequate land, through land improvement and reform, is a precondition for the future of common field agriculture.

The upper elevation limit of common field agriculture is frequently dependent on the territorial delimitation of the sector. In actual practice, sectoral boundaries of common field agriculture are often designed to abut adjacent communities or to grade into nonarable upper elevation sites. The importance of development change at the upper elevation limit is due to the active disintensification process that is occurring at the least productive, upper elevation margins of mountain agriculture (Zimmerer, 2003a). Indeed, the number and area of fields at upper elevations in the Cuzco and Cochabamba regions have diminished by as much as 40% during the past 20–30 years due to concomitant changes that include labor-time reallocation to off-farm work and intensified midelevation agriculture that includes marketing potatoes and agribusiness-based contract growing of malt barley. This decrease appears less marked, however, in those communities with common field agriculture, such as Mollomarca. For this reason, common field agriculture holds the possibility of considerable promise for community- and multicommunity-based resource management and agrobiodiversity conservation (Pestalozzi, 2000; Zimmerer, 2002b).

Informal-style coordination of cropping and livestock raising is also a rationale for patchiness. By choosing to plant crop types with similar growing calendars and rotation patterns, the farmers of adjacent parcels tried to reduce the extent and occurrence of livestock damage. The benefits of this crop–livestock coordination were referred to by 11 of the farmers interviewed in the Mollomarca community ($n = 15$) and by 10 cultivators that were interviewed in the Pampa Churigua community ($n = 29$). Still another factor in favor of patchiness stemmed from the shared tendency to plant a single crop in a particular agronomic habitat, such as an irrigated area. This third rationale was mentioned by 14 Mollomarca growers, who referred specifically to irrigated patches of off-season potatoes, and by four farmers in the Pampa Churigua area.

MODELS FOR LANDSCAPES

The models of inhabitants for their mountain landscapes are important cultural constructs. Their cultural models are described below as representations of space (Lefebvre, 1974/1991). The model expressed by the 15 farmers in the Mollomarca community is a framework that consists of four agricultural landscape units to which all their fields belong. Persons identified the three units as "valley" (*kheshwar*), "hill" (*loma*), and "early planting" (*maway*) (Table 7.4). The fourth unit was referred to as either "oxen area" (*yunglla* or *yuñlla*; derived from *yunta*, or "ox team") or "midslope" (*chawpi qhata*). Six farmers claimed that this latter space was unnamed, although evidently different. Farm units were named and distinguished on the basis of topography and production traits, principally crop types, calendar, tillage, social relations of production, and yield disposition. Similarly, all farmers of the Pampa Churigua community shared a general model for their agricultural landscape. Their model consists of three units that encompassed all fields within the community. The units are distinguished on the basis of production traits and named, according to topography, as "valley floor" (*pampapi*), "slope" (*faldapi*), and "hilltop" (*patapi* or *puntapi*) (Table 7.5).

Referring to these cultural expressions, I used transect walks to estimate the locations of each farm unit within the agricultural landscapes of the two communities. This research was undertaken in conjunction with the agro-

TABLE 7.4. Major Farm Units (Socioenvironmental) of Mollomarca Community (Cuzco, Peru)

Farm unit	Production traits	Social relations	Disposition of harvest	Existing range (meters)	Ideal range (meters)	Spatial cohesion
"Valley" (*qheshwar*)	Maize and intercrops (quinoa, tarwi, beans, squash)	Household (especially women) and local markets	Consumption, commerce, seed	3,100–3,550	3,100–3,550	Moderate
"Hill" (*loma*)	Andean potato and tuber crops in raised beds (*wachus*)	Household and local markets	Consumption, seed, commerce	3,800–4,100	3,550–4,000	High
"Oxen area" (*yunlla*)	Tuber crops and grains other than maize	Oxen tillage, agribusiness contracting, expanding, urban market	Household, commerce, consumption	3,000–3,800		Low–Moderate
"Early planting" (*maway*)	Off-season potato planting (irrigated)	Household, expanding off-season market	Commerce	3,000–3,600		Moderate

TABLE 7.5. Major Farm Units (Socioenvironmental) of Pampa Churigua Community (Cochabamba, Bolivia)

Farm space	Main crop(s)	Social relations	Disposition of yield	Existing range (meters)	Ideal range (meters)	Spatial cohesion
"Valley floor" (*pampapi*)	Maize	Household and local markets	Commerce, consumption, seed	2,800–3,200	2,800–3,000	Low
"Slope" (*faldapi*)	Wheat	Household and local markets	Commerce, consumption, seed	2,800–3,450	3,100–3,350	Low
"Hilltop" (*patapi*)	Potatoes	Household and local markets	Commerce, consumption, seed	2,800–3,450	3,350–3,450	Low

ecological transect methods described above. Each of the transects was walked twice (with a different community member each time). The key informants were asked to name the local farm unit(s) at each interval of 100 meters of elevation. Existing locations of each farm unit were found to cover a wide elevation range (Tables 7.4 and 7.5). Cultural farm spaces overlapped considerably. The extent of overlap varied, however, among units. Range of the "hill" unit, for example, demonstrated less overlap and sharper boundaries. Cohesion of this space was due both to the environmental limits of a cool climate (that prohibits cultivation of many crops) and to the spatial practice of common field agriculture (described above in the preceding section).

Cartographic study and map-based visualization techniques were also used. The broad range and substantial overlap of cultural units posed a special cartographic challenge in representing the agricultural landscapes of the study communities. A map design was innovated in order to display the existing location of farm units while it also illustrates overlap (Figures 7.2 and 7.3). For each farm unit, a fill pattern is made up of one background shade and one marking. Where the farm units overlap, the background is chosen to represent the predominant type, while markings indicate the overlap of secondary spaces. This map design offers a contrast to standard illustrations of mountain land use, for the usual illustrations of tiered zonation models depict a stacked, layer-cake-like arrangement of discrete zones (Buzhuo et al., 1997; Kappelle & Juárez, 1995).

Cultural models of the people in the Mollomarca and Pampa Churigua communities create a paradox, however, when compared to the overlapping patchwork-like patterns of the actual landscape. The immediate source of the paradox is that certain names or terms that are commonly applied to the spatial units of the farm landscape are derived from topographical features that suggest the role of tier-like areas (valley bottom, valley, slope, hill and hilltop; Tables 7.4 and 7.5). Paradoxically, their terms for these units seem to contradict the actual pattern of overlapping patchworks. The suggestion of zones in their cultural models is not, however, specific or direct. People in the Mollomarca and Pampa Churigua communities, like their counterparts in other Peruvian and Bolivian places, tend to conceive of landscape space as unbounded and relational. Therefore a "lower" field can often be located above a "higher" one. Thus the tiered quality that is suggested by the names of their farm units must be seen as a general landscape guide or ideal.

Their reliance on topographic terms in local landscape models may be useful insofar as it offers an ideal standard that can serve as a basis for comparison to current reality and to possible future changes. Interviews, participant observation, and historical research indicate that topographic names are used to emphasize the distinctness of landscape areas. Distinctness aids the people of Mollomarca and Pampa Churigua as a means of exerting land rights and expressing cultural ideals. The topographic basis of naming in the cultural landscape model thus helps politically to delimit, protect, and reclaim land rights. Ideally, each family in the communities holds access to all major areas of farming. People frequently buttress their land claims by referring to topographic terms (e.g., "valley" fields) that delimit their notion of accustomed or proper access.

A second ideal is a cultural goal related to food security that is symbolized by the topographic names for landscape areas. Farmers in Mollomarca and Pampa Churigua believe that the "hill" and "valley" units designated in the cultural model ought to be producing areas of potatoes and maize, respectively. Ideally, these twin foodstuffs are the principal staples of local eating. (In actual fact, everyday diets include a sizeable share of other foodstuffs, especially inexpensive wheat products like imported macaroni noodles and rice.)

In addition, a cultural model of broader economic ideals is also suggested by the landscape representations of the Mollomarca and Pampa Churigua people. Name meanings and shared symbols weld the subsistence-provisioning functions of "hill" and "valley" units to the market-centered ideals of the "oxen area" and "early planting." The latter areas are represented as the chief allotments to marketing. (Names of the latter derive from aspects of tillage and the growing calendar, respectively, that underlie commercial advantage.) Ideally, the overall landscape of locally preferred development combines secure self-provisioning spaces ("hill" and "valley") with advantageous marketing units ("oxen area" and "early planting").

The ideal development model differs, however, from existing landscape usages in the communities of Mollomarca and Pampa Churigua. Notably, the disposition of harvest is not a diametric contrast between self-provisioning and marketing, as suggested by the ideal, but rather these allocations are mixed in each farm space (Tables 7.4 and 7.5). Findings of overlapping patchworks, along with cultural models that incorporate zone-like terminology, direct attention to the historical formation of the landscape practices and patterns.

Historical evidence indicates that local models for the landscape have been prone to alterations as a result of environmental, socioeconomic, and political forces. This evidence suggests that overlapping patchworks did not originate *de novo* as a result of recent development changes. Farm spaces occupied by the "early planting" and "oxen area" took shape as features of the post-Conquest landscapes that were driven by economic and political policies of the Spanish colonial empire (Gade, 1992). Colonial estate fields identified with the "early planting" and "oxen area" were long interspersed in the landscapes of eastern Cuzco, and thus overlapped with "valley"-type fields, some of which belonged to peasant tenants (Archivo Departmental de Cusco, 1784–1785, 1794–1796). Indeed, farm landscapes in this region were shaped through integration with "global" forces associated with Spanish colonialism beginning in the 1500s and, still earlier, by extraregional integration with the Inca state.

Common field agriculture, perhaps the most notable element of the local models for landscape, is decidedly nonstatic and shows no intrinsic incompatibility with development change. Indeed, this cornerstone of the local model for landscape bears an overall resemblance to other malleable arrangements of resource use that have undergone the dynamic evolution of their meanings and environmental characteristics (broadly analogous to the *sasi* of Indonesia; see Zerner, 1994, 2000). Prone to adjustment, the common field agriculture of the Andes is also analogous to the particular features of the canal-based irrigation cluster, or *suyu*, that is a fundamental spatial unit of much grassroots or alternative development in the Andes today (Zimmerer, 2000a, 2000b). (As a result, common field agriculture should be thought of as the potential basis of a useful environmental planning territory.)

DISCUSSION: A MODEL
OF OVERLAPPING PATCHWORKS

The findings of this study suggest that a pattern of overlapping patchworks is common to the mountain agriculture of communities in Cuzco, Peru, and Cochabamba, Bolivia. (The term "overlapping patchworks" is chosen since it does not indicate the regularity of variation that would be inferred by a descriptor such as "mosaic." Also, overlapping patchworks is intended to evoke a spatial image that is missing from a generic term such as "configuration.") Overlapping patchworks of mountain agriculture are shown to be shaped through the political–ecological interaction of farm livelihoods and mountain environmental factors. Key proximate conditions in this interaction are high variability and low predictability of agroenvironmental parameters (rainfall, landslides and geomorphic disturbances, pest and disease outbreaks) and the ecological versatility of adaptive traits of the major crops (potatoes, barley). Also important to crop location decisions are multifaceted cropping ratio-nales and crop–livestock coordination that arise through the direct influence of extraregional processes. Since the patterning and processes of overlap-ping patchworks are common in certain regions, and perhaps in particular the eastern Andes where compressed zonation is common, this political–ecological model is proposed as a regionally specific counterpart of the tier-based zonation model.

This discussion draws attention to the ecological dynamics of the politi-cal–ecological model of overlapping patchworks that is suggested. In ecologi-cal terms, the landscape-shaping role of mountain farm environments can dif-fer from the uniformity and predictability of the model of tier-based zonation. Key environmental factors show moderate–high degrees of spatial variability and temporal unpredictability in the Cuzco and Cochabamba landscapes. An-nual rainfall, mean temperature, and frost risk tend to display high variability and low predictability in regions of the eastern and central Andean regions in general (Knapp, 1991; Winterhalder, 1994). Microclimate factors tend also to exhibit similar properties. High variability and low predictability of climate parameters may be increased according to possible scenarios of climate change associated with global warming in tropical mountain environments such as the Andes.

High variability and low predictability are also produced through the irregular occurrence of landslides, slumps, and debris flows that affect the Cuzco and Cochabamba landscapes. Geomorphic disturbances directly affect agricultural soil properties such as depth, quality, and available moisture (Zimmerer, 2002c, 2003b). Perceptions of these soil traits influence the deci-sions of farmers about where to locate fields of certain crops based on the cri-terion of good yield. Crop pest and disease outbreaks at the landscape scale also are characterized by high variability and low predictability. Like soils-shaping geomorphic disturbances, pest and disease outbreaks influence farm-ers' perceptions of field suitability. Crop pest and disease outbreaks in tropical mountain regions also are likely to increase under possible scenarios of global climate change (Price & Barry, 1997).

Environmental tolerances of a moderate–high level are displayed in the basic agroecological properties of several of the major food plant types in these mountainous farm landscapes. Potatoes and barley in particular exhibit a high degree of ecological versatility of adaptive traits (i.e., broad genotype-environment interaction) with respect to the elevation-related variation of Andean environments. The agroecological versatility of the Andean potato crop is dependent upon metapopulation processes of gene flow and seed dis-persal. Collectively, the suite of ecological tendencies discussed above (high variability/low predictability, ecological versatility, and adaptive dynamics) suggest how certain key processes that are characteristic of tropical mountain landscapes display a tendency toward nonequilibrium properties.

In political and economic terms, the overlapping patchworks differ sharply from the suggestions of spatial segregation, specialization, and relative stability that are typically portrayed in the environmental zonation model of land use. Dynamic local variability of farm production and land use is customary of contemporary Andean farming land use. These changes tend to be highly localized, and frequently include concomitant intensification and disintensification in different places within community boundaries. Specific immediate conditions leading to agricultural change in the Cuzco and Cochabamba areas are intensification of potato production (in midelevation areas due to off-season growing), agribusiness-contracted farming of malting barley, and extensive off-farm labor migration. The growing complexity and volatility of rural economies coincides with macroeconomic restructuring in Peru and Bolivia (Reardon, Berdegué, & Escobar, 2001).

Overall, this study's proposal of a model of overlapping patchworks suggests the prominence of the following groups of key processes: (1) the spatial variability and temporal unpredictability of biogeophysical factors; (2) the adaptive dynamics and environmental tolerances of crop types; (3) sociospatial incentives (e.g., crop–livestock coordination) and disincentives (on-farm labor-time shortage) that impinge directly on the capacity of farmers to coordinate land use and to create landscape cohesion; and (4) spatial and temporal variability of farm input and product markets exacerbated by fluctuating regional development, nongovernmental organization projects, and governmental policies. Change-prone states and interaction within this group of factors tend to ensure overlapping patchworks in the Cuzco and Cochabamba landscapes.

CONCLUSION:
IMPLICATIONS FOR CONSERVATION MANAGEMENT
AND DEVELOPMENT FOR SUSTAINABILITY

The dynamics of overlapping patchworks in particular places, such as the areas of this study, may offer an alternative to the scaling of crop areas through tier-based zonation. "Overlapping" describes the fact that such areas often are partially superimposed through the increasingly diversified production styles of small- and medium-size resource users in developing countries—and that they are changeable in extent. "Patchworks" refers to the internal heterogeneity that is characteristic of areas of land and resource use. These units are perceptually distinct; they are labeled and commonly referred to with terms that refer to a mixture of natural and social features (reflecting nature–society "hybrids"; see Swyngedouw, Chapter 5, this volume); and they display substantial within-unit mixtures of resource practices. The idea of overlapping patchworks may offer a reasonable alternative to overly rigid definitions and static assumptions of conservation management units that derive from the model of tiered zonation.

One general implication for attempts to combine conservation management with development is that they may face the likelihood of greater flexibility and less certainty in the spatial–environmental properties of mountain agriculture. Sizeable flexibility is evidenced by landscape-related resource use in Andean regions such as Cuzco and Cochabamba. Functioning and location of farm spaces arranged in overlapping patchworks evidence more flexibility than is posited in the tiered zonation model. Concomitantly, less certainty is also evident. The functioning of overlapping patchworks renews focus on the interconnected spatial, environmental, and political–ecological qualities of mountain agriculture land use. By contrast, the model of tiered zonation can be used to reduce such qualities to purely environmental factors. That latter style of reasoning would result in the deduction of spatial forms from basic assumptions that are proven unfounded in the case studies of this research.

A specific implication regards the integration of diverse Andean food plants (such as Andean potatoes, maize, quinoa, and ulluco) into policies, programs, and projects for the *in situ* conservation of agrobiodiversity that is coupled with development gains for the growers themselves. These food plants are valued by farmers, including many Cuzco and Cochabamba people, for reasons of agroecology, taste, nutrition, and culture (such as identity formation). Their capacity to grow the diverse food plants amid current development change depends on whether the farmers can cultivate both the diverse crop types and commercial high-yield varieties (HYVs). Such capacity hinges, *inter alia*, on the spatial–environmental parameters of this mountain agriculture.

Overlapping patchworks are a sign of moderate–high flexibility in the cultivation of several food plants. *In situ* conservation is possible if farmers are able to adjust and retain their cultivation of diverse food plants. In Mollomarca, for example, farmers have combined the diverse food plants, which distinguish the "hill" and "valley" units, with the cropping of HYVs that are less diverse and that are sown in the "ox area" and "early planting" spaces. By contrast, farmers of some nearby communities in eastern Cuzco (e.g., Colquepata, Chocopía) have replaced their "valley" space, and its richness of agrodiversity, with enlargements of the "oxen area" and "early planting." Thus the potential contribution of spatial–environmental flexibility to agrobiodiversity conservation may ultimately be marked by a significant degree of uncertainty. Future research should investigate the uses and limits of landscape flexibility for conservation-compatible outcomes.

REFERENCES

Archivo Departamental del Cusco. (1784–1785, 1794–1796). Legajo 89, Causas Ordinarias, Intendencia. Cuzco, Peru.

Bassett, T. J., & Zimmerer, K. S. (2003). Cultural and political ecology. In C. Wilmott & G. Gaile (Eds.), *Geography in America, 2000* (pp. 282–307). Oxford, UK: Oxford University Press.

Bernard, H. R. (1994). *Research methods in anthropology: Qualitative and quantitative approaches.* Thousand Oaks, CA: Sage.

Blaikie, P. (1994). *Political ecology in the 1990s: An evolving view of nature and society* (CASID Distinguished Speaker Series, No. 13). East Lansing: Michigan State University.

Blaikie, P., & Brookfield, H. (Eds.). (1987). *Land degradation and society.* London: Methuen.

Brookfield, H., & Padoch, C. (1994). Appreciating agrodiversity: The dynamism and diversity of indigenous farming practices. *Environment, 36,* 6–11, 37–45.

Brush, S. B. (1976). Man's use of an Andean ecosystem. *Human Ecology, 4*(2), 147–166.

Brush, S. B. (1995). *In situ* conservation of landraces in centers of crop diversity. *Crop Science, 35,* 346–354.

Buzhuo, P., Lijie, P., Haosheng, B., & Higgitt, D. L. (1997). Vertical zonation of landscape in Tibet, China. *Mountain Research and Development, 17*(1), 43–48.

Denevan, W. M. (2001). *Cultivated landscapes of native Amazonia and the Andes.* Oxford, UK: Oxford University Press.

Gade, D. W. (1992). Landscape, system, and identity in the post-Conquest Andes. *Annals of the Association of American Geographers, 82,* 461–477.

Kappelle, M., & Juárez, M. E. (1995). Agroecological zonation along an altitudinal gradient in the montane belt of the Los Santos Forest Reserve in Costa Rica. *Mountain Research and Development, 15*(1), 19–37.

Knapp, G. (1991). *Andean ecology: Adaptive dynamics in Ecuador.* Boulder, CO: Westview Press.

Lauer, W. (1993). Human development and environment in the Andes: A geoecological overview. *Mountain Research and Development, 13*(2), 157–166.

Lefebvre, H. (1974/1991). *The production of space* (D. N. Smith, Trans.). London: Blackwell.

Mayer, E. (1985). Production zones. In S. Masuda, I. Shimada, & C. Morris (Eds.), *Andean ecology and civilization* (pp. 45–84). Tokyo: University of Tokyo Press.

Mayer, E. (2002). *The articulated peasant: Household economies in the Andes.* Boulder, CO: Westview Press.

Painter, M. (1995). Upland–lowland production linkages and land degradation in Bolivia. In M. Painter & W. H. Durham (Eds.), *The social causes of environmental destruction in Latin America* (pp. 133–168). Ann Arbor: University of Michigan Press.

Pestalozzi, H. (2000). Sectoral fallow systems and the management of soil fertility: The rationality of indigenous knowledge in the High Andes of Bolivia. *Mountain Research and Development, 20*(1), 64–71.

Price, M. F., & Barry, R. G. (1997). Climate change. In B. Messerli & J. D. Ives (Eds.), *Mountains of the world: A global priority* (pp. 409–445). New York: Partenon.

Reardon, T., Berdegué, J., & Escobar. G. (2001). Rural nonfarm employment and incomes in Latin America. *World Development, 29*(3), 395–409.

Rodriguez, L. O., & Young, K. R. (2000). Biological diversity of Peru: Determining priority areas for conservation. *Ambio, 29*(6), 329–337.

Sarmiento, F. O. (2000). Breaking mountain paradigms: Ecological effects of human impacts in managed Tropandean landscapes. *Ambio, 29*(7), 423–431.

Stevens, S. (Ed.). (1997). *Conservation through survival: Indigenous peoples and protected areas.* Washington, DC: Island Press.

Troll, C. (1968). The cordilleras of the tropical Americas: Aspects of climatic, phytogeographical and agrarian ecology. In *Geo-ecology of the Mountainous Regions of the Tropical Americas* (pp. 15–56). Bonn: Ferd Dummlers Verlag.

Wells, M. P., & Brandon, K. E. (1993). The principles and practice of buffer zones and local participation in biodiversity conservation. *Ambio, 22*(2–3), 157–162.

Western, D., & Wright, R. M. (Eds.). (1994). *Natural connections: Perspectives in community-based conservation.* Washington, DC: Island Press.

Winterhalder, B. (1994). The ecological basis of water management in the central Andes. In W. P. Mitchell & D. Guillet (Eds.), *Irrigation at high altitudes: The social organization of water control systems in the Andes* (pp. 21–67). Washington, DC: American Anthropolical Association.

Zerner, C. (1994). Through a green lens: Customary environmental law and community in Indonesia's Maluku Islands. *Law and Society Review, 28*, 1079–1122.

Zerner, C. (Ed.). (2000). *People, plants, and justice: The politics of nature conservation.* New York: Columbia University Press.

Zimmerer, K. S. (2000a). Reworking conservation: Nonequilibrium landscapes and nature–society hybrids. *Annals of the Association of American Geographers, 90*(2), 356–369.

Zimmerer, K. S. (2000b). Re-scaling irrigation in Latin America: The cultural images and political ecology of water resources. *Ecumene, 7*(2), 150–175.

Zimmerer, K. S. (2002a). Agrodiversidad de las communidades de montaña. In F. Sarmiento (Ed.), *Montañas del mundo: Una prioridad global con perspectives latinoamericana* (pp. 429–443). Quito, Ecuador: Abya-Yala.

Zimmerer, K. S. (2002b). Common field agriculture as a cultural landscape of Latin America: Development and history in the geographical customs of land use. *Journal of Cultural Geography, 19*(2), 37–63.

Zimmerer, K. S. (2002c). Social and agroenvironmental variability of seed production and the potential collaborative breeding of potatoes in the Andean countries. In D. A. Cleveland & D. Soleri (Eds.), *Farmers, scientists, and plant breeding: Integrating knowledge and practice* (pp. 83–107). Wallingford, UK: CABI International.

Zimmerer, K. S. (2003a). Geographies of seed networks and approaches to agrobiodiversity conservation. *Society and Natural Resources, 16*(3).

Zimmerer, K. S. (2003b). Just small potatoes: The use of seed-size variation in "native commercialized" agriculture and agribiodiversity conservation among Peruvian farmers. *Agriculture and Human Values, 20,* 107–123.

Zimmerer, K. S. (2003c). *Globalization and the new geographies of conservation.* Unpublished manuscript, Department of Geography, University of Wisconsin—Madison.

Zimmerer, K. S., & E. D. Carter. (2002). Conservation and sustainability in Latin America. *Yearbook of the Conference of Latin Americanist Geographers, 27,* 22–43.

CHAPTER 8

Environmental Science and Social Causation in the Analysis of Sahelian Pastoralism

Matthew Turner

Livestock grazing has consistently been implicated in the degradation of African drylands (Middleton & Thomas, 1997; Sinclair & Frywell, 1985). Despite long-lasting concern, our understanding of the biophysical and social processes that contribute to grazing-induced ecological deterioration remains limited. Resource scientists/managers have typically measured the potential for "overgrazing" by comparing the livestock density (stocking rate) in a given area to an estimated carrying capacity. Local stocking rates are seen to fluctuate largely through livestock demography (birth and death rates). Undue reliance on this seemingly straightforward measure is both symptom and cause of continued ignorance of the effect of grazing on Africa's savanna and steppe environments. By presenting a political ecological analysis of pastoral management of cattle in central Mali (West Africa), this chapter provides a more spatially and socially informed environmental analysis of livestock population dynamics and grazing-induced environmental change in dryland areas.

Political ecology introduces two major perspectives much needed for understanding the relationship between livestock and grazing-induced environmental change in the Sahel. The first is political ecology's attention to scale—both the spatial and the organizational (social and ecological) scales of indige-

Adapted by Matthew Turner from the article "Overstocking the Range: A Critical Analysis of the Environmental Science of Sahelian Pastoralism," published in *Economic Geography,* 69(4), 402–421 (1993). Adapted by permission of Clark University. The adapting author is solely responsible for all changes in substance, context, and emphasis.

nous management of grazing animals, vegetative production, and range assessment tools. To study the effect of livestock activities (trampling, grazing, defecating) on rangeland ecology, one needs to directly tie these activities to a particular piece of ground through which ecological processes occur (see Bassett & Koli, Chapter 6, this volume). This is especially important in the dryland tropics where changes in soil or vegetation have often been misdiagnosed as caused by livestock when in fact they may be caused by climate fluctuations. Livestock actions affect different ecological processes at particular spatial and temporal scales. An explicit recognition of the multiscaled effects of livestock activities is necessary to determine how livestock actions have influenced land productivity.

The integration of the social and ecological processes is the second major contribution of political ecology to environmental analysis of livestock-induced environmental change. Livestock are not only biological populations but economic units—they are raised by humans as stores of wealth and as commodities (like food plants; see Zimmerer, Chapter 7, this volume). Therefore, fluctuations in local livestock populations are driven not only by an area's rangeland ecology (affecting livestock fodder and disease vectors) but also by its economy. A more spatially explicit environmental assessment of livestock grazing requires an understanding of what may affect not only the size of an area's livestock population but its spatial distribution within the area. Especially in the open range situations that prevail in much of dryland Africa, the spatial distribution of livestock is determined in large part by the land use and herding decisions of people. Where people choose to graze their livestock is determined not only by an area's fodder resources and animal species (goats, sheep, cattle, camels, etc.), but by a myriad of other factors including the need of the herding family to be near cropped fields or milk markets and the experience and energy devoted by "the herder" to herding the animals. Therefore social factors beyond those associated with the livestock/fodder balance affect the density of grazing pressure across agropastoral landscapes. Decisions made by herders, farmers, and livestock owners are affected by ecological conditions, and these conditions may change due to the interaction of past decisions and rainfall. The tight interconnections among the "ecological" and the "social" requires an integrated approach such as that offered by political ecology. The integration offered by political ecology is achieved through careful attention to the structured complexity of socioecological interactions.

ENVIRONMENTAL ASSESSMENT AND THE SIMPLIFICATION OF ECOLOGICAL COMPLEXITY

The Sahelian region of West Africa (Sahel) is the strip of land lying south of the Sahara Desert that receives an average of 100–600 millimeters of rain each year, sporadically falling from June through September. As in other dryland areas of Africa, vegetative characteristics (production, cover, nutritional content, species composition) fluctuate widely from year to year, driven in large part by rainfall fluctuations (Penning de Vries & Djitèye, 1982). Directional change, leading to the persistent decline in vegetative productivity for a defined purpose, is often referred to as "land degradation."[1] A myriad of biophysical pathways exist through which livestock actions (e.g., trampling, excretion, defoliation) could lower the availability of plant growth factors (soil fertility, soil structure, seed stock, photosynthetic tissue, moisture), and in so doing lower the productive potential of the land. These pathways work at different spatial and temporal (time) scales. For example, the active removal of plant tissue by grazing animals can lower the productivity of a grazed patch (around 10–20 square meters) during the same growing season. By contrast, livestock grazing for many decades can lead to the redistribution of plant nutrients over broader spatial scales (50–100 square kilometers)—lowering nutrient availability in grazed areas and raising it at resting/watering points (Turner, 1998). The relative importance of these pathways in affecting vegetative productivity during any particular growing season depends on soil type, rainfall, local topography, and the timing and intensity of grazing.

The study of degradation pathways is complicated by the high spatial and interannual variation in vegetation caused by differences in rainfall. Not only does this variability complicate the search for trends and for causal relationships, the effect of past and present livestock actions on vegetative characteristics in any given growing season is strongly influenced by that season's rainfall (Penning de Vries & Djitèye, 1982; Turner, 1998). Therefore, signs of "grazing impact" are far from obvious and the identification of grazing-induced degradation requires ecological work that pays careful attention to degradative pathways and ecological processes while controlling for rainfall variability.

Unfortunately, the basic components of experimental research to account for rainfall variability (experimental controls, replications, and statistical significance testing) have not been utilized in much of the ecological work in the Sahel. Rather, this work has been dominated by range inventory and mapping studies, whose goal is to estimate the amount of palatable forage available to local livestock populations (Boudet, 1972, 1975; Le Houérou, 1989). This information is used to estimate a stocking rate that the inventoried range can support—that is, a carrying capacity (Behnke, Scoones, & Kerven, 1993).

Implicit in such approaches is the assumption that, despite high natural variation in rainfall, changes in stocking rate have a discernable effect on the condition of the range. This reflects the influence of equilibrium views of range ecology on both the European phytosociology and North American range management traditions (Stoddart, Smith, & Box, 1975). For both, rangeland, in the absence of grazing and drought, is seen to evolve along a single continuum toward a climax state (the succession–climax model; see, e.g., Clements, 1916). Increased grazing pressure or drought leads to poorer range condition, while reduced grazing pressure or high rainfall improves range condition. Conceptualized in this way, range condition during droughts can be improved by reducing grazing pressure (Westoby, Walker, & Noy-Meir, 1989).[2]

The applicability of the succession–climax model to arid and semi-arid rangelands has been seriously questioned (Behnke et al., 1993; Ellis & Swift, 1988; Turner, 1998; Westoby et al., 1989). It fails to distinguish ephemeral from persistent change in rangeland production. Moreover, it fails to distinguish the effects of grazing from the effects of climate. Even so, it has had a significant influence on the agenda and methodologies used in Sahelian range monitoring and management. Within the short-term, crisis-oriented funding environment for scientific work in the Sahel, the model is attractive because it provides easy management rules and does not require the significant allocation of time and effort that is necessary to diagnose degradation (e.g., persistent declines in productivity), identify its causes, and understand how local people manage their livestock across dryland landscapes. Pasture of poor condition, no matter what the cause, should be managed in such a way as to improve its productivity. One cannot control climate, but it may be possible to control livestock populations. By such reasoning, regulating livestock populations becomes the major management and policy prescription offered by range experts.

Complex ecology, highly fluctuating rainfall conditions, short-term funding, and the conceptual models underlying Western range management all contribute in different ways to the limited experimental study of grazing ecology in the Sahel. As a result, researchers have surprisingly little to say about the importance of grazing, let alone the biophysical pathways by which livestock actions affect vegetative productivity in the Sahel. In the absence of useful data, stocking rates are used as proxies for ecological stress. From field observations, perceived spatial relationships between "range condition" and stocking rate are used as evidence for, or against, grazing-induced degradation. Given the high spatial variability of rainfall, rainfall and grazing effects are hopelessly entangled in such perceptions.

TRUNCATION OF CAUSAL LINKAGES

The dominance of stocking rate as the major analytic variable to measure ecological stress is both cause and symptom of scientists' limited knowledge of the biophysical processes leading to grazing-induced ecological degradation. Under conditions of recurrent drought in the African dryland regions, management prescriptions have been simply to destock. By adopting an overly simplistic view of rangeland ecology for developmentalist expediency, resource management experts fail to study particular degradation pathways that can be tied to changes in livestock population and their grazing management. In so doing, they truncate the political–ecological linkages that run through livestock management and unknowingly produce a very simple model of the rural societies that own and manage the livestock in question. It is only through the delineation of linkages between political economy, management, and ecology that a much wider array of environmental policy initiatives can be discovered.

The carrying-capacity approach to range management in the Sahel views problems associated with grazing as resulting from the imbalance between livestock population and range resources. Dynamic characterizations of this population/resource balance most often treat the increase and decrease of livestock numbers as solely a question of demography—dependent solely on the availability of fodder and disease vectors. This ignores the fact that these livestock populations are also economic entities—they are managed by their owners as producers of milk or traction, as commodities, and as stores of wealth. Most all rural peoples in dryland Africa seek to store their wealth in the form of livestock due to the liquidity, growth, and mobility advantages of livestock. Sahelian West Africa has historically supplied urban markets to the south with meat (Kervan, 1992). Studies have found that most livestock owners sell a large fraction of the annual growth of their herds (Amanor, 1995; Sutter, 1987; Wagenaar, Diallo, & Sayers, 1986, p. 12). Therefore, slight variations in offtake significantly affect growth rates of individual herds. In addition, offtake rates vary with respect to the wealth of the owner, with richer owners marketing a lower fraction of their herd growth (Sutter, 1987; Turner, 1992, pp. 162–163). The growth of cattle populations within rural regions of the Sahel is therefore a result of the interplay between the extraregional demand for meat, reflected in the prices cattle merchants are willing to pay, and local demand for cattle as stores of wealth and productive capital.[3] These demands are produced not simply from the production of economic surplus but from the distribution of that surplus within local political economies. Such a perspective provides the basis for a fuller characterization of the confluence of social and biophysical factors working to change local livestock populations.

Stocking rate is commonly used to measure the ecological pressure produced by a livestock population. Unfortunately, estimates of an area's livestock population by themselves provide little ecologically significant information. Such information needs to be coupled with knowledge of how animals are distributed within the area, which necessarily requires a greater understanding of local grazing management. Stocking rate is equal to the average number of animals grazing a rangeland over a defined time period divided by the surface area of the rangeland. Certainly, stocking rate has ecological relevance—changes in the number of animals at a particular location (range site) during a particular season is useful information for environmental evaluation. However, the ecological relevance of a stocking rate value declines when it is averaged over broader temporal and spatial scales. For example, an annual stocking rate averaged over a broad geographic region provides very little ecologically significant information. Even for livestock inventories conducted at finer spatial scales, the seasonality of the spatial distribution of livestock within the inventoried perimeter is as ecologically important as the overall stocking rate.[4] Rangelands in dryland Africa are largely unfenced, and livestock, due to the heterogeneous and ephemeral nature of fodder resources, are managed in a mobile fashion. The changing spatial distribution of livestock during the year is therefore determined in large measure by the management of daily and seasonal grazing movements by livestock owners and herders. The management of grazing movements, in turn, is strongly affected by changing patterns of access to pasture resources, labor resources, and livestock among livestock-rearing households. In this respect, changes in grazing movements are intimately tied to changes in the broader political economy.

Adopting a more complex view of rangeland ecology demands that environmental analysis moves away from static comparisons of average stocking rates with carrying capacities toward approaches that incorporate a greater understanding of changes in local livestock populations and their seasonal distribution across rangelands. Such approaches will necessarily require a greater understanding of local management decisions and how they are shaped by the broader political economy. Figure 8.1 represents a political ecological framework for the environmental analysis of agropastoralism in dryland Africa. A number of key features of this framework can be highlighted. First, given the contingent nature of biophysical and social processes and their interaction, one can only meaningfully understand socioecological processes leading to land degradation for a clearly identified piece of land (Blaikie, 1985). In other words, the analysis needs to be spatially explicit. Such an analysis allows one to identify the managers whose livestock utilize the area and whose decisions affect the number of animals and their management and distribution within the area. By tying social institutions, people, and their animals to land, political ecology allows for greater political–ecological integration. By extending lines of causation through management practices to the broader political economy, political ecological analyses can identify a wider range of policy initiatives influencing grazing patterns than can conventional environmental assessment.

In the following section, a case study of the political ecology of livestock populations and grazing management in the Maasina region of central Mali is presented. A political ecology approach similar to that presented in Figure 8.1 was followed. A portion of the western rain-fed border of the Maasina floodplain, an area previously identified as "overgrazed," was chosen for ecological work (Figure 8.2). Ecological research was performed within this study area to delineate important degradative pathways associated with livestock activities. Results of this research are not presented here, but a major finding was that the productivity of annual grasslands in the study area is particularly sensitive to heavy grazing pressure during the middle of the rainy season (July–August).[5]

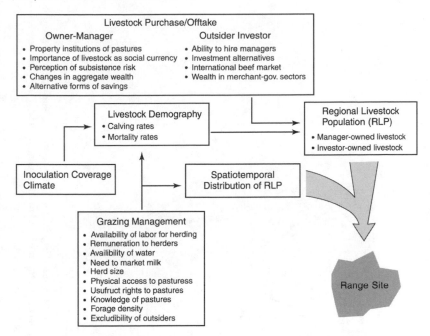

FIGURE 8.1. The seasonal magnitude of grazing pressure at a range site (with dimensions of 1–25 hectares, depending on the patchiness of vegetation) is determined by the regional livestock population and its spatiotemporal distribution. Each in turn is affected by both ecological and social variables. The regional livestock population is affected by management, climate, and inoculation coverage through livestock demography, as well as by factors affecting the buying and selling of livestock by people with rights to pasture in the area encompassing the range site. Outside investors often view livestock as one of a number of possible investments and entrust their livestock to livestock managers who value livestock for their productive capacities as well. Whether the range site experiences ecologically significant grazing pressure from the regional livestock population is determined by the management of livestock grazing movements by livestock managers and reflects not only the ecological characteristics of the site and its location within a broader biogeographic context but the degree to which managers have access to labor, knowledge, livestock, market, and pasture resources.

The pastoral clan whose cattle herds move through the study area were the focus of more qualitative ethnographic research. The case study traces how changes in the local political economy contributed to the rapid growth of cattle populations in the 1950s and 1960s. In addition, it will be shown how changes beginning during the 1970s have led to grazing management that has increased the presence of herds during the middle rainy season, when the range is most sensitive. The case study concludes by considering the policy implications of the findings of this research.

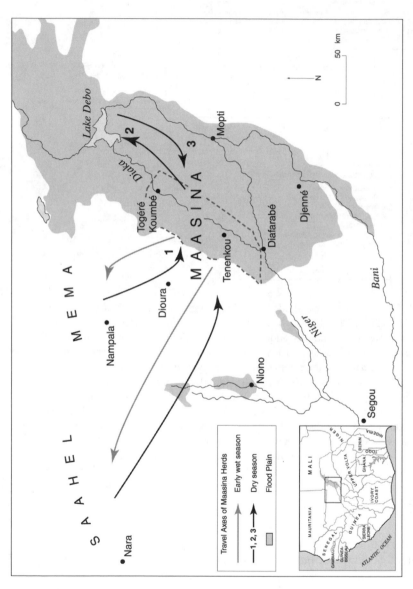

FIGURE 8.2. The location of present-day Maasina on the floodplain (shaded areas) of the inland Niger delta of the Republic of Mali. Pastoral herds converge on the western border of the Maasina floodplain at the end of the rainy season and move into the interior of the floodplain as the dry season progresses—only leaving the floodplain for dryland pastures to the northwest with the first rains. Travel axes of Maasina transhumance herds during the early wet season, the beginning of the dry season (1), the early–mid-dry season (2), and the end of the dry season (3) are designated by arrows. Important rainy season regions for Maasina herds, the Saahel and the Mema, are also designated.

CASE STUDY AREA

The Inland Niger Delta (the Delta) is a 20,000-square-kilometer floodplain fed by the Niger and Bani Rivers in the Sahelian zone of Mali (Figure 8.2). It is one of the most important livestock and rice-growing centers of West Africa. The Delta region receives sparse and variable rainfall from June through September. Long-term average annual rainfall varies from 400 to 650 millimeters, increasing from north to south. This places it squarely within the southern portion of the Sahelian bioclimatic zone. The Delta's location within such a zone has made it an important multipurpose resource for many different occupational and ethnic groups. Fisherfolk, rice cultivators, and cattle herders all utilize the floodplain during a portion of the year.

Rice cultivators constitute the majority of the population. The major rice-growing ethnic groups are the Marka (Nono), Songhay, and, most importantly, the Rimayβe, former slaves of the Delta Fulβe. The dominant form of rice cultivation in the Delta is performed without local flood control or use of fertilizers and pesticides. Plowing is by animal traction and hoe. Yields are variable, but generally quite low.

Herds managed by Fulβe (Fulani, Peul, Fula, and others), Kel Tamashek (Touareg), Bella, and Maure groups use the Delta floodplain as dry season pasture (November–June). These herds are largely composed of cattle, but sheep, goats, and donkeys are not uncommon. The Delta represents one of the most important dry season pastures in Africa, with one of the highest regional cattle densities in West Africa (International Livestock Centre for Africa, 1981, p. 5; Resource Inventory and Management Limited, 1987, pp. 69–71). Approximately 1.5 million cattle converge on the floodplain at the end of the rainy season, moving toward its more deeply flooded areas as the dry season progresses (Figure 8.2). Pastures of the Delta are controlled by the Delta Fulβe clans, who grant access to outsiders.

The pastoral economy of the Delta has long been oriented toward trade, with foodstuffs supplied to desert-side market towns through the Jenne–Timbuctu trade route. The proximity of dry season pastures to these regional trade centers has led to the integration of local livestock markets into regional trading networks. A web of cattle markets located near dry season pastures supplies regional markets. Over the past 40 years, estimates for cattle herds in the Delta consistently find herd offtakes to be around 12% (Coulomb, 1972; Gallais, 1967, pp. 410–411; Wagenaar et al., 1986, p. 12).

Prior to colonial conquest, local political and military power in the Maasina had been held by Fulβe clans for close to 500 years. The productivity of a complex slave economy allowed the Fulβe elite to support large classes of artisans, griots, and Islamic clergy.[6] Early in the colonial period, French-brokered negotiations led to an agreement between Maasina slaves and the Fulβe elite that abolished many servile obligations. Former slaves (Rimayβe) gained greater control over land and rice, and through this control became major agents of livestock investment in the Maasina by the mid-20th century.

DISTRIBUTION OF CATTLE WEALTH

Cattle represent the major store of wealth in the Maasina. Virtually all cattle owned in the Maasina are found in Fulβe-managed herds. A survey of the 18 herds managed by the Fulβe clan with whom I worked (2,000 cattle enumerated) reveals that 86% of the cattle were not owned by the managing family but entrusted to them by other people. Most of the existing cattle derive from old investments—investments made prior to 1973, which marks the beginning of a 25-year period of recurrent drought and stagnation of the Maasina economy. Only those who have obtained surplus from outside the region (labor migrants) or who have extracted surplus from a large group of rural producers (government officials and merchants) made cattle investments during the economically troubled 1972–1989 period. Cattle wealth is fairly well distributed across major social groups: Rimayβe (54% of the local adult population), while historically not cattle owners, own 34% of surveyed cattle. Livestock-rearing Fulβe do not own a disproportionate share of cattle. Fulβe-owned cattle (entrusted and self-owned) represent 32% of all surveyed cattle, which is only slightly higher than the fraction of Fulβe in the local population (25% of the adults in the surveyed clan's village).

While cattle wealth is distributed fairly evenly across social groups, it is unevenly distributed within the social groups of the Maasina. At least 48% of the cattle population is owned by 10% of cattle owners, a group that represents only a small part of the total population. A small number of Islamic clergy, government officials, and merchants own 70% of the entrusted cattle not owned by herding Fulβe or Rimayβe (representing 30% of all entrusted cattle). It is within these groups that disproportionate amounts of wealth have accumulated.

Despite the growing importance of merchant/government official ownership, cattle ownership today is much more widely distributed across all social groups than it was in the precolonial period, when the noble Fulβe and Islamic clergy were the major cattle owners. Historical reconstructions of inheritances within this Fulβe clan show that much of their cattle wealth was lost before the 1970s, with losses over the whole period attributed equally to cattle deaths and sales to buy grain. This suggests that significant shifts in the ownership of the Maasina cattle population occurred prior to, or concurrent with, the rapid growth of cattle populations during the 1960s. A more detailed historical analysis of wealth transfers and accumulation is required to determine whether the expansion of the cattle population was due in part to changes in the allocation of surplus accumulation.

GROWTH OF THE MAASINA CATTLE POPULATION (1950–1970)

The cattle population in the Sahel expanded tremendously during the 1950s and 1960s. Estimates of aggregate cattle population in four Sahelian countries (Senegal, Mali, Burkina-Faso, Niger) increased from 1950 to 1968 at an average annual rate of 11% per year (Le Houérou, 1989, p. 126).[7] The 1971–1973 drought ended this period of growth, with loss estimates for Sahelian countries ranging from 20 to 40% (Toulmin, 1985). Similar oscillatory patterns in livestock populations are observed in other arid and semi-arid rangelands of Africa. On the surface, such fluctuations resemble those of some wild animal populations, prompting many analysts to characterize domestic livestock populations in Africa as controlled simply by livestock demography, range productivity, and climatic factors.

As with other parts of the Sahel, the cattle population in the Maasina expanded rapidly from the mid-1950s through the 1960s. This resulted in a local stocking rate viewed as excessive by range ecologists (Boudet, 1972; International Livestock Centre for Africa, 1981). Although there are differences in opinion about the long-term ecological consequences of the high stocking rates of the early 1970s (Gallais, 1984; Turner, 1992, pp. 331–398), it does represent a period of unusually high ecological stress due to grazing. With recurrent drought and economic decline, the regional cattle population has not subsequently approached these levels.

The expansion of the cattle population and the high grazing pressure of the early 1970s cannot be explained solely by reduced mortality resulting from successful inoculation coverage and abundant rains. Using evidence from existing ownership distributions, reconstructions of cattle wealth histories of local informants, and the written observations of colonial officials, I argue that the expansion of cattle populations would not have been as great if not for substantial changes in resource access and surplus control within Maasina society.

The historiography of cattle ownership and accumulation is extremely difficult for the Sahel given the lack of quantitative data concerning cattle populations, trade flows, and cattle ownership. While cattle population estimates for the Maasina are thought to be more accurate than elsewhere in the Sahel, there are few reliable data concerning trade flows and cattle ownership. Anecdotal information gleaned from historical documents, along with knowledge of precolonial and present-day distributions of ownership, are therefore used to reconstruct the history of local transfers of cattle wealth in the Maasina and to evaluate their effect on the observed growth of cattle populations from 1950 to 1970.

Prior to the 1940s, investment in cattle by Rimayβe and caste groups was very limited. The Fulβe and clergy elite still controlled the vast majority of Maasina cattle. Despite the successful cattle inoculation programs of the colonial Service de l'Élevage, which began during the 1930s and eventually reached most Maasina herds by 1945, there was no net growth in the cattle population from 1930 to 1951 (Figure 8.3). The lack of growth was due in large measure to colonial confiscation of cattle that led to considerable losses of Fulβe-owned cattle, especially during World Wars I and II (Drahon, 1949; Gallais, 1967, p. 396). Inhabitants of the Maasina also attribute the lack of growth in cattle populations during the early colonial period to lack of interest in cattle accumulation (*hakille jow'di*) among landholding Fulβe clans. These clans used a substantial portion of the surplus extracted from their slaves and from outside clans seeking access to floodplain pastures to support the livelihood of a dependent group of griots, Islamic clergy, weavers, silversmiths, and goldsmiths.

Initial cattle investments by Rimayβe were spurred by their rapid adoption of the plow (Figure 8.3). By 1957, close to half of all Rimayβe families in the Delta owned at least one plow (Mission Socio-Économique du Soudan, 1961, p. 10)—a remarkable statistic given that as late as 1934 rice cultivators were reportedly resisting the adoption of the plow (Gallais, 1967, p. 232). Greater access to land and increased control over agricultural surplus gave the Rimayβe the economic means and incentive to purchase plows and cattle. In the late 1950s, Rimayβe made up 46% of the Maasina population, whereas

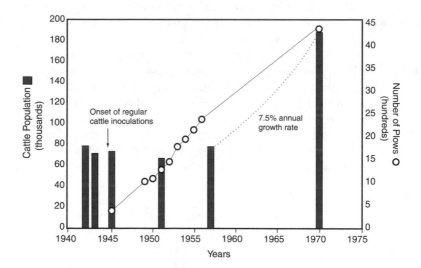

FIGURE 8.3. Growth in the numbers of cattle and plows in the Maasina (1940–1970). *Sources*: Cattle population estimates were adapted from Drahon (1949) for 1942 and 1943; Gallais (1967, p. 395) for 1945, 1951, and 1957; and Coulomb (1972, p. 59) for 1970. Numbers of plows in 1945–1956 were derived from Gallais (1967, p. 232). The number of plows in 1970 were estimated from the total for the Delta multiplied by the 1956 ratio of Maasina/ Delta plows (Gallais, 1984, p. 188).

Fulβe, Islamic clergy, and artisan caste groups together represented 30% of the population (Mission Socio-Économique du Soudan, 1963). Even modest levels of per capita accumulation of cattle among Rimayβe would therefore have had a significant effect on local stocking rates. The rapid conversion of Maasina agriculture to animal traction, along with growing evidence for accumulation of cattle among nonherding groups, occurred at the onset of a period of remarkable growth in the Maasina cattle population (Figure 8.3). This expansion occurred during a period of relatively limited growth of human populations in the Maasina, with annual growth rates over the period ranging from 0.7 to 0.9% (Gallais, 1984, 185; Turner, 1992, pp. 45–52). Even after taking into account certain deficiencies in the data, the 7.5% average annual growth rate of the cattle population in the Maasina between 1957 and 1971 cannot be explained by a reduction of mortality due to increased inoculation coverage nor to an increase in fecundity. The higher rate of regional cattle growth must in part be due to a lower rate of net exports from the region.

Anecdotal evidence suggests that the local demand for cattle was quite high during the 1950s and 1960s, with local prices exceeding those prevalent in the export markets. For example, a local administrator observing the Tenenkou cattle market in 1955 notes that "price levels at the first Tenenkou markets are higher than those last year. A steer weighing 350–375 kilograms sells for 15,000–16,000 francs. These prices are somewhat inexplicable since the cattle merchant himself states that the cattle prices in the Ivory Coast and the Gold Coast are very low" (my translation; Revue Trimestrielle, 2/20/1955, Cercle de Macina I-E-40 FR, Archives Nationales du Mali). A significant part of these intraregional exchanges were from herding Fulβe to other groups (Gallais, 1967, p. 409).

In sum, historical changes in the distribution of cattle ownership within the Maasina reflect the changing structure of the regional economy. A hierarchical class structure, determined largely by ethnicity and caste, evolved toward a capitalist system in which ethnicity plays a lesser role in determining the control of surplus. Of critical importance was the changing control over agricultural surplus from Fulβe nobles to former slaves during the first half of the colonial period. This trend accelerated after independence, as the state more actively supported the land claims of rice cultivators against the Fulβe elite. As Rimayβe payments to their former Fulβe masters became increasingly symbolic, new accumulation strategies of the more numerous Rimayβe influenced the livestock sector. Agricultural surplus was freed from being used to maintain and gain access to land. Surplus control shifted from the Fulβe elite, who spent a large fraction on supporting their griots, clerics, and artisans, to a larger number of Rimayβe, whose moral economy discouraged conspicuous consumption. Increasing individualism and material accumulation among the Rimayβe led to a growing interest in covert accumulation in the form of cattle.

The lack of serious droughts and epidemics during the 1960s created the biological conditions necessary for the rapid expansion of the Maasina cattle population. The evidence presented above supports the argument that changes in the local patterns of accumulation led to an increase in local demand for cattle as stores of wealth and productive capital relative to the effective demand for meat in export zones. This resulted in a greater retention of cattle in the Maasina than would be expected from regional market conditions alone. The cattle boom in the Maasina during the 1960s cannot be viewed solely as a result of biological procreation but is the result of the confluence of productive as well as social factors.

CHANGES IN GRAZING MANAGEMENT (1970–1995)

Recurrent drought since the early 1970s has led to a decline in the Maasina cattle population and a reduction in the number of animals that could influence vegetation along the western border of the floodplain. The degree to which cattle population declines have resulted in a reduction in ecologically significant grazing pressure in the study area depends on how cattle are managed (Figure 8.1). Changes in the travel movements of herds to and from the floodplain could affect how long the herds stay along its dry border during the beginning and end of the rainy seasons. Moreover, changes in daily grazing movements from encampment points within the dry border region affects the density of grazing pressure produced by a resident cattle population. Less dispersal away from encampment and watering points leads to higher grazing densities than would be predicted by regional averages.

While drought has led to a reduction in the cattle population, changes in management caused by the combination of drought-induced impoverishment of herd managers, water shortage, and the shifts in livestock wealth that preceded the drought have together led to a situation in which grazing management has worked to exacerbate the grazing pressures felt on border pastures. I will summarize these changes, which are more fully described and analyzed elsewhere (Turner, 1999). Recurrent drought has led to reduced flooding in the Maasina along with slowly developing pastures in the beginning and earlier drying of watering points at the end of the rainy season in the vast pastures to the northwest. In addition, rebel activity led to increased security threats to herds at these rainy season pasture areas during the 1990s. These changes have made it both physically more difficult and more risky to herd animals in the traditional pastures to the northwest and easier to remain on or near the Maasina floodplain during the rainy season. Given the strong pride in herding and the negative stigma attached to late departures or early returns to the floodplain among Maasina herders, these changes alone have not led to widespread disregard for traditional rainy season transhumance to the northwest. It has been herds managed by families that are either livestock-poor, resulting in small herds or herds dominated by entrusted cattle, or labor-deficient, that have remained longer along the dry border at the beginning and end of the rainy season. In extreme cases, herds will simply remain in the dry border region throughout the rainy season. Given that the annual grasslands there are particularly sensitive to heavy grazing pressure during the middle of the rainy season, these changes are of significant environmental concern.

The impoverishment of herding families, leading to the allocation of family labor away from herding and an increased reliance on cattle entrustments from investors, has reduced the incentives and capacities of herders to herd diligently on a daily basis. Entrustment contracts provide very little remuneration to herders, and families must often rely on young boys to herd 10–16 hours a day. Interviews of herders along with livestock productivity data support the conclusion that this has resulted in an increased concentration of livestock around encampment and watering points in the dry border region. Increased aggregation in the grazing patterns within the region increases the ecological stress produced by a given cattle population.

While regional cattle populations have declined from the levels reached in the late 1960s, ecological and social conditions have led to grazing management practices producing greater ecological impact on the western dry border per head of cattle in the Maasina. In this way, two countervailing trends were experienced during the 1970–1995 period with respect to the environment impact of the Maasina cattle economy. There is no reason to believe that such changes will always work in opposite directions to buffer the livestock grazing system. Improvements in rainfall coupled with reduced exports of Maasina cattle will likely lead to greater local cattle populations. Such changes are not likely to be associated with an improvement in their grazing management. Cattle wealth will remain in the hands of nonmanagers. Herding families will benefit somewhat from the increased productivity of their patrons' wealth but will remain impoverished in cattle well into any recovery cycle. Under these conditions, large livestock populations combined with poor management will dramatically increase the environmental threat posed by Maasina cattle husbandry.

CONCLUSIONS

I have examined problems associated with the use of stocking rate as the major analytic variable to evaluate the ecological sustainability of animal husbandry in the Sahel. In so doing, I have shown that environmental analysts have often utilized conceptual models of ecology and society that do not match the reality of the Sahel. These models tend to "biologize" the etiology of grazing-induced degradation by ignoring, or greatly simplifying, the social processes that can affect the timing and intensity of grazing pressure.

Through critical examination of such models, I have argued for a more complex causal model of grazing-induced degradation, in which social processes are neither ignored nor simply invoked as broad forces, but instead are explicitly tied to grazing ecology. Two mediating variables affect the timing and intensity of grazing pressure within a range area: livestock population and temporal/spatial distribution of livestock. Each in turn is affected by both biophysical and social factors (Figure 8.1).

Lengthening the chains of causation beyond the livestock–forage balance toward grazing management and the human decisions that lead to livestock population growth opens up whole new areas of environmental policy. Environmental policy and management is no longer simply focused on ways to elicit destocking by force or market incentives. Instead, the policy initiatives concerned with livestock population are likely to be concerned with trade, the investment decisions made by the urban-based investor, and the distribution of livestock wealth. The Sahelian cattle industry now competes in urban beef markets to the south that have become increasingly global over the past 25 years. Beef/cattle trade policies, promulgated by West African states, Argentina, or the European Economic Community will, by affecting exports from the Sahelian zone, affect the size of local livestock populations. Urban-based investors, despite having access to a range of investment opportunities, often continue to amass huge numbers of cattle that are entrusted to many different herders. Why? Livestock ownership is not a particularly lucrative economic activity. This is an area that is likely to be amenable to policy interventions that provide alternative investment opportunities that match the security, liquidity, and secrecy needs of such investors.

This chapter has emphasized the environmental importance of grazing management. The environmental impact of a given livestock population is strongly mediated by how that population is managed across dryland landscapes. The carrying capacity model of environmental assessment virtually ignored the role of grazing management. In so doing, herders were typically seen as tradition-bound peoples who were solely interested in livestock accumulation and, as such, portrayed as ecological villains (Sinclair & Frywell, 1985). Three points can be made in response to this common perception. First, very few managers of livestock today have the economic freedom to treat livestock as simply prestige items even if they may have had such freedom in the distant past. Second, livestock managers and herders are often not the owners of the livestock in their herds. Therefore, one should not conflate livestock management with livestock ownership. Third, the management of livestock toward productive ends (not simply investment goals) in such technology-scarce, dryland environments does not necessarily conflict with that which would be prescribed to protect rangeland productivity. Both seek to maximize the distribution of livestock to match livestock to available forage resources.

These responses suggest a set of policies that do not attack the herding profession but actually are designed to support managers' production interest to maximize the dispersion of herds across the forage resource. Policies that support the mobility of herds are quite important in this regard. In a Sahelian context, such policies would involve the enforcement of existing laws that were passed to protect key pastoral sites from agricultural encroachment (water points and transhumance corridors). In addition, policies and development initiatives that work to improve the economic security of livestock managers are still likely to lead to greater investments of labor into grazing management, given the persistent cultural connections to the profession. Focused on stocking rates, resource management and development experts have been hesitant to support the livestock profession since it was felt that development aid largesse would simply be converted into cattle, thereby exacerbating the environmental situation. Certainly, issues of livestock investment need to be scrutinized (as mentioned above), but policies directed at lowering investment pressure in livestock should be different from those designed in support of proper grazing management. The vast majority of livestock in a herd are owned by others. Small increases in livestock owned by the herding family will not only increase its economic security but increase its incentives to properly manage the whole herd.

A key area in need of policy innovation is in the area of the entrustment/herding contract that exists between owners and herders. In the Maasina, there exists a situation where all recognize that the remuneration received by herders for their herding services is insufficient. Herders state that they must steal entrusted livestock in order to support themselves, while owners state that they are unwilling to provide more remuneration because the herders steal. The level of distrust between the two groups has led in many cases to significant social cleavages within Maasina communities. This standoff can only be broken by increasing the security of the entrustment contract for both parties. This could be done by a combination of improved ownership marking of cattle and a brokered agreement concerning herder remuneration involving government officials, Islamic clergy, livestock merchants, and herder organizations.

Given the ephemeral nature of resource availability and the mobility of production systems in the dryland Africa, policy initiatives must be multipronged and work at different spatial and social organizational scales. This will necessarily involve both community-based initiatives (e.g., to resolve tenure and entrustment contract disputes) and state-led initiatives (such as trade, investment, and transhumance corridor protection). Political ecology, with its analytical emphasis on political–ecological integration and attention to spatial and organizational scales, is ideally suited to develop such combined policy initiatives.

NOTES

1. The degree to which the manager is interested in promoting beef production, milk production, soil conservation, crop production, or the viability of local wildlife populations for an area will affect how environmental change is interpreted. In a range context, the biological characteristics of most concern include the above-ground biomass, forage quality, and species composition of lignaceous (trees and shrubs) and herbaceous (grasses and forbs) vegetation.
2. Such management prescriptions are consistent with commonsense notions that vegetation during a drought is more sensitive to grazing pressure. Certainly there is less vegetation during drought years, and therefore a greater percentage of vegetation may be consumed and livestock nutrition is likely to suffer. But what about the sensitivity of pastures to grazing-induced degradation? Results from range ecological research suggest that rangeland productivity is more sensitive to grazing during the rainy season and that a number of degradation pathways are likely to be more active during wetter years (Turner, 1998).
3. The "interplay" should not viewed as a simple market equilibrium. Livestock markets are imperfect: these markets are highly volatile, buyers and sellers lack full information, the export trade is heavily regulated, and merchants display cartel-like behavior.
4. Typically, stocking rates are estimated at scales for which livestock population data are available: annual estimates for the whole administrative district. Due to their high level of spatial and temporal aggregation, such estimates supply very little ecologically relevant information.
5. The results of this research are reported elsewhere. Changes in soil structure, species composition, same-season production, and soil fertility, while working at different spatiotemporal scales, were all found to be most often associated with livestock grazing during the middle period of the rainy season (Turner, 1998).
6. Griots in the Maasina are often members of artisan castes who provide various services to nobles, including recounting and praising noble lineages, storytelling, resolving disputes, and facilitating business transactions.
7. Such estimates are suspect and may be overestimates due to early underreporting of livestock numbers. Taking this into account, the growth rate may be closer to 8%, which is still a remarkably high rate of increase.

REFERENCES

Amanor, K. S. (1995). Dynamics of herd structures and herding strategies in West Africa: A study of market integration and ecological adaptation. *Africa, 65*(3), 351–394.

Behnke, R. H., Scoones, I., & Kerven, C. (Eds.). (1993). *Range ecology at disequilibrium.* London: Overseas Development Institute.

Blaikie, P. (1985). *The political economy of soil erosion in developing countries.* London and New York: Longman.

Boudet, G. (1972). *Projet de développement de l'élevage dans la région de Mopti* (Étude

Agrostologique No. 37). Maisons-Alfort, France: Institut d'Élevage et de Médecine Vétérinaire des Pays Tropicaux.

Boudet, G. (1975). *Manuel sur les pâturages tropicaux et les cultures fourragères*. Paris: Institut d'Élevage et de Médecine Vétérinaire des Pays Tropicaux.

Clements, F. E. (1916). *Plant succession: An analysis of the development of vegetation*. Washington, DC: Carnegie Institute.

Coulomb, J. (1972). *Projet de développement de l'élevage dans la Region de Mopti: Étude du troupeau*. Maisons Alfort, France: Institut d'Élevage et de Médecine Vétérinaire des Pays Tropicaux.

Drahon, M. (1949). Notes sur un recensement du cheptel bovin du Diaka. *Bulletin des Services de l'Élevage et des Industries Animales de l'Afrique Occidentale Français, 1–3*, 19–24.

Ellis, J. E., & Swift, D. M. (1988). Stability of African pastoral ecosystems: Alternate paradigms and implications for development. *Journal of Range Management, 41*, 450–459.

Gallais, J. (1967). *Le Delta intérieur du Niger*. Dakar, Senegal: IFAN.

Gallais, J. (1984). *Hommes du Sahel: Espaces-temps et pouvoirs. Le Delta intérieur du Niger, 1960–1980*. Paris: Flammarion.

International Livestock Centre for Africa. (1981). *Systems research in the arid zones of Mali* (ILCA Systems Study No. 5). Addis Ababa, Ethiopia: International Livestock Centre for Africa.

Kervan, C. (1992). *Customary commerce: A historical reassessment of pastoral livestock marketing in Africa* (ODI Agricultural Occasional Paper No. 15). London: Overseas Development Institute.

Le Houérou, H. N. (1989). *The grazing land ecosystems of the African Sahel*. Berlin and New York: Springer-Verlag.

Middleton, N., & Thomas, D. (Eds.). (1997). *World atlas of desertification*. New York: UNEP.

Mission Socio-Économique du Soudan. (1961). *Enquête budgétaire dans le Delta central nigérien*. Paris: Service de Coopération de l'Institut National de Statistique et des Études Économiques.

Mission Socio-Économique du Soudan. (1963). *Étude démographique du Delta vif du Niger*. Paris: Service de Coopération de l'Institut National de Statistique et des Études Économiques.

Penning de Vries, F. W. T., & Djitèye, M. A. (1982). *La productivité des pâturages sahéliens*. Wageningen, The Netherlands: Centre for Agricultural Publishing and Documentation.

Resource Inventory and Management Limited. (1987). *Refuge in the Sahel*. St. Helier, UK: Resource Inventory and Management Limited.

Sinclair, A. R. E., & Frywell, J. M. (1985). The Sahel of Africa: Ecology of a disaster. *Canadian Journal of Zoology, 63*, 987–994.

Stoddart, L. A., Smith, A. D., & Box, T. W. (1975). *Range management* (3rd ed.). New York: McGraw-Hill.

Sutter, J. W. (1987). Cattle and inequality: Herd size differences and pastoral production among the Fulani of northeastern Senegal. *Africa, 57*(2), 196–218.

Toulmin, C. (1985). *Livestock losses and post-drought rehabilitation in sub-Saharan Af-*

rica (LPU Working Paper No. 9). Addis Ababa, Ethiopia: International Livestock Centre for Africa.

Turner, M. D. (1992). *Living on the edge: Fulβe herding practices and the relationship between economy and ecology in the Inland Niger Delta of Mali.* Unpublished doctoral dissertation, University of California, Berkeley.

Turner, M. D. (1998). The interaction of grazing history with rainfall and its influence on annual rangeland dynamics in the Sahel. In K. Zimmerer & K. Young (Eds.), *Nature's geography: New lessons for conservation in developing countries* (pp. 237–261). Madison: University of Wisconsin Press.

Turner, M. D. (1999). Labor process and the environment: The effects of labor availability and compensation on the quality of herding in the Sahel. *Human Ecology, 27*(2), 267–296.

Wagenaar, K. T., Diallo, A., & Sayers, A. R. (1986). *Productivity of transhumant Fulani cattle in the Inner Niger Delta of Mali* (ILCA Research Report No. 13). Addis Ababa, Ethiopia: International Livestock Centre for Africa.

Westoby, M., Walker, B., & Noy-Meir, I. (1989). Opportunistic management for rangelands not at equilibrium. *Journal of Range Management, 42*(4), 266–273.

PART IV

GEOSPATIAL TECHNOLOGIES AND KNOWLEDGES

CHAPTER 9

Fixed Categories
in a Portable Landscape

The Causes and Consequences
of Land Cover Categorization

Paul Robbins

Satellite imagery and other forms of remotely sensed data have been introduced into environmental debates to settle them and to clarify the trajectory of environmental change with reference to hard facts. Yet such imagery,
rather than reducing the contentiousness of landscape change claims, actually
reinforces it. One satellite image, for example, that I recently took on a circu-
itous tour through a small town in India, evoked myriad interpretations.
Going door-to-door and visiting along the way a number of local inhabitants,
I observed a wide range of interpretations of the same hard data. Foresters
pored over the image and traced their fingers along the edges of a rugged line
of hills, suggesting that the cover there represented evidence of reforestation.
Farmers pointed to bare soils and showed denuded areas where forest had dis-
appeared on the edge of villages. A retired forester decried the disappearance
of tree cover along the forest fringe. A worker at a local advocacy organization
for pastoralists identified large swaths of grassland lost to trees. The same frag-

Adapted by Paul Robbins from the article "Fixed Categories in a Portable Landscape: The Causes and
Consequences of Land-Cover Categorization," published in *Environment and Planning A, 33*, 161–179
(2001). Adapted by permission of Pion Limited, London. The adapting author is solely responsible for
all changes in substance, context, and emphasis.

ile satellite image supported the tremendous weight of many interpretations, some of them complementary, others contradictory.

Such a range of interpretations should be in no way surprising. Landscapes are constantly read and reread in the practice of daily life, and the plurality of interpretations underlines the instability of any single reading (Barnes & Duncan, 1992). This is especially true in a world of proliferating landscape images, including traditional maps, air photography, and satellite imagery. The mobility of these technologies and technological artifacts, evidenced by the presence of satellite images and geographic information systems (GIS) platforms in remote villages, puts more landscapes into the hands of more people. Landscapes are therefore highly *portable* and are increasingly mobile with every technological innovation in geographic science.

Even so, in the act of bounding, naming, and describing important patterns in a satellite image, including deforestation, reforestation, and desiccation, observers must set the categories of analysis to convey the urgency of real-world phenomena. Land cover changes in these complex landscapes can take form only by *fixing* the categories of their interpretation—for example, one must identify forest in order to map deforestation.

The questions that emerge in the wake of this understanding of the relationship between geography and epistemology are threefold. First, since maps of land cover derived from local knowledge are rarely analyzed through controlled comparison, it remains unclear whether the landscape accounts of environmental professionals and other groups are consistently divergent (Robbins, 2000). Do they differ in a uniform fashion, and along what axes of difference? Second, the degree to which emergent technologies, like remote sensing and GIS, have effected public and professional landscape conceptions is also underexplored. Has the advent of these portable technologies changed the perceptual apparatus of development professionals and local people? Third, and finally, the relationship between these emergent technologies and actual landscape change is also largely unknown. Has the measurement of the landscape through remote platforms led to new kinds of landscapes?

Using a participatory mapping technique, I here explore these questions, examining not only the causes of clashing land use classifications, but also their effects. The method employed satellite imagery to map the competing landscape conceptions of professional foresters and local producers in a case study from the region surrounding the Kumbhalgarh Wildlife Sanctuary in Rajasthan, India. Starting from land cover definitions elicited from photograph identification and interviews with herders, farmers, and forestry officials, I conducted a classification of the same multispectral data sets twice, based on the divergent views of expert foresters and local land users. The resulting digital images were used to compare the differing categories of land cover both *qualitatively*, in terms of their meaning, and *quantitatively*, in terms of their spatial coverage and location.

The analysis suggests that competing maps of the region share common ground in some areas of the landscape but are marked by profound contradictions in others. In particular, the willingness of foresters to accept as "forest" the expanding savanna scrublands dominated by invasive species stands in marked contrast to the views of locals, who often see such landscapes as degraded. In the process, incentives are formed for professional forestry to reproduce these landscapes of invasives. Thus, an ecomanagerial bureaucratic imaginary at Kumbhalgarh enables the parameters of remotely sensed data to reengineer the landscape. More generally, the results of the study underline the fact that satellite imagery is not an impartial tool for the settlement of debates about land cover but is instead the result of prior debates about the character of nature.

These results suggest the profound integration, indeed the inextricable linkage, of social/political and ecosystem dynamics revealed in political ecological analysis. Trees grow through their relationship both to statistical techniques and to forest bureaucracies, while institutions and measurements are influenced by the aggressive growth of trees. In this way, landscapes are engineered to suit measurement, rather than the other way around. The implications for understanding geospatial technologies and ecological processes extend well beyond the Indian context.

CATEGORIES IN A COMPLEX LANDSCAPE

The arrival of satellite imagery to help interpret landscape change is nowhere more happily anticipated than in the Godwar region of Rajasthan, adjacent to the Kumbhalgarh Wildlife Reserve, flanking the Aravalli hills of the southern Pali district (Figure 9.1). The area is a semi-arid farming belt adjacent to a hilly deciduous forest dominated by perennially green deciduous xerophytic tree species. The predominant mode of subsistence is a mixed agropastoral strategy that depends upon both intensified production techniques and a heavy use of the wild resources of forest and fallow land. As a result foresters and local nongovernmental organizations (NGOs) both clamor for comprehensive data on the state of land resources, especially forest, since sweeping policy changes in recent decades have made spatial and temporal management decisions imperative. The Rajasthan Forest Act of 1955, which empowered the enclosure of the Kumbhalgarh Reserve, allows a great deal of discretionary room for varying management strategies and practices. Foresters and locals must negotiate when and where to enclose sections of the forest from grazing, who should be allowed access to forest resources, and where to focus Forest Department efforts with regard to environmental decline and its amelioration.

In particular, it is crucial to determine exactly how much forest there is, where the forest is, and whether it is in expansion or decline, since the limited re-

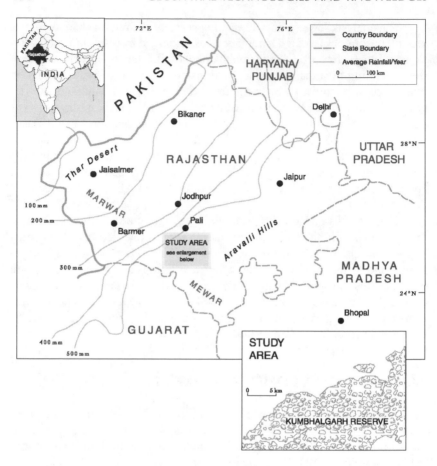

FIGURE 9.1. The Kumbhalgarh Wildlife Sanctuary, located on the spine of the Arvalli Hills in southern Rajasthan, India.

sources of the forest management bureaucracy must be spread over a wide array of management tasks. The topographical and ecological complexity of the region makes such an assessment difficult. The reserve, though not overlarge, is thickly wooded and difficult to penetrate. The adjacent plains are covered with a mixed savanna that supplies timber, fodder, and fuel resources for villagers, but which varies tremendously in species mixes, density, and rates of decline and recovery. So too there is disagreement about the rate and location of forest cover change. Local people insist that forest cover is in decline, but the location of that change and its driving causes vary from individual to individual and group to group (Robbins, 2000). Foresters, on the other hand, largely agree that forest decline has halted in the last decade and that recovery is in evidence in some places.

Thus almost all parties greet the prospect of satellite image information and mapping tools with interest. Foresters are sure that it will verify their claims that forest cover is stable or expanding. Locals believe it will reveal the decline in key resources. Problems remain, however, for implementing satellite image analysis to address these questions. Specifically, the image classification process requires that definitions for key features of the landscape be known before the fact (Robbins & Maddock, 2001). However, the complex landscape vocabulary of local people and its divergence from that of professional foresters makes establishing these claims categories difficult. Where definitions of land cover types differ between foresters and locals, as is often the case, it is necessary either to choose one interpretation over the other or to accept divergent classifications of reflectance "reality." Of the two options, the latter method, drawing as it does on participatory GIS techniques (Weiner, Warner, Harris, & Levin, 1995) and countermapping (Peluso, 1995), provides a better window into the role and effect of remote sensing technology on the landscape.

METHOD: PARTICIPATORY CLASSIFICATION TECHNIQUE

The research was conducted using field-based interviews and observation, joined with remote sensing techniques and GIS-based classification analysis. SPOT image satellite scenes of the study region were obtained and 27 sites were selected to represent a range of vegetation mixes and land uses. These were photographed and the ground cover at each location was recorded. Photos of each of these sites were given to 68 local producers and nine foresters for identification. The sample was purposive, representing producers across a range of caste (*jati*) communities and land endowments, and included nine women. Participants were asked to identify the photographs, providing whatever land cover category they believed each photograph represented. For the purposes of this study, aimed at exploring relationships between local and professional knowledge, each site was given two identification categories, the first based on the category given to the site by a majority of locals, and the second based on the identification used by professional foresters.[1] Using the photos, these were then classified into two separate land cover maps, one based on producer categories and aggregations, the other based on those of professional foresters.

Of six pictures of mixed scrub, for example, foresters might identify three as "forest." The reflectance characteristics of the three areas corresponding to those photos would then be used to generalize the "forest" coverage for the entire region. The resulting map would show the distribution of all areas that foresters would likely identify as "forest." Locals, on the other hand, might identify three entirely different photos as "*junglat*" (or forest), in addition to two or three others that foresters identified as something else. The local image would show a very different distribution of "*junglat*," therefore, representing only those areas that locals would consider to be forest. The resulting images were compared quantitatively and qualitatively.[2]

COMPETING CATEGORIES OF LAND COVER

The differences between local and professional categories are notable in several regards (Table 9.1). First, locals reported a much larger number of categories overall and forester categories were more uniform. This reflects the relatively more uniform typology of land covers institutionalized through forester experience and training. Foresters, particularly those in the middle ranks, describe with pride their training in forestry schools like the Forest Research Institute in Dehra Dun and their retraining in workshops occasionally held in this or other official centers. At these sites, categories are established and reinforced and the naming of landscapes is institutionalized. For foresters without such formal training, repeated daily interaction drives and unifies category sets; to socially succeed as a forester requires the use and deployment of uniform forester knowledge. This, in turn, reflects the self-organizing tendencies of institutions. Institutions do the naming for individuals; state forestry does the categorizing for foresters.

Second, forester categories were notable in that they were predominantly given in English (Tables 9.1 and 9.2), whereas local categories were provided either in Hindi (as taught in state grade schools) or in Marwari (a local dialect). In the specific categories of state forestry, there remains a strong influence of colonial knowledge. Specifically, the categories of foresters reflect the categories of the census system, relict in the region from the imposition of survey techniques of the British political agent when the Rajputana states were put

TABLE 9.1. Collapse of Categories for Comparison

Producer category	Includes	Forester category	Includes
Junglat	*Jangal*; *kharva*; *magara*	Forest	Forest
Kheti	"Farm"; *erat*; *baghicha*; *Kheti*	Cultivated	Agricultural; *khetadari*; orchard
Gav/banjar	*Aakariya*; *bakhar*; *banjar*; *kharas*; *medan*; *partal*; *tutal zamin*; *vala*	Urban/bare	*Abadi*; *nadi*; town
Akad/gocher	*Gocher*; *jordh*; *reliya ki zamin*; *reveni*; *simada*	Grassy/fallow	Fallow; revenue waste

TABLE 9.2. Comparison of Coverage Based on Producer and Forester Categories

Category	Producer Coverage	Forester Coverage	Agreement (K)
Junglat or Forest	328.5	444.1	.34
Kheti or Cultivated	111.2	103.3	.97
Banjar or Bare	91.1	34.6	.63
Gocher or Fallow	132.3	100.9	.86

under colonial residency in the 19th century (Rajputana-Gazetteers, 1908). Most prominently, the colonial-era category of "waste" land persists from that time, and continues to include all land "suitable for cultivation" that is not already under the plow. These lands are valuable resource areas for locals who use them for grazing, fodder, and food collection. The instrumental character of forester categories is therefore notable, if in no way surprising.

Local categories are notable in that they show an extreme sensitivity to processes of ecological succession and change. By distinguishing length of fallow, reduction in productivity, and density of growth, locally defined categories are not only spatial, they are *temporal*, and reflect sophisticated notions of ecogenic and athropogenic change. The length of fallowing time, for example, distinguishes *jordh* and *simada* lands, while *banjar* and *partal* land are distinguished by varying forces of degradation and change. This mirrors the characteristics of indigenous land cover classifications more generally, which commonly recognize "continua, successions, tendencies, and cycles" (Ellen, 1982, p. 223).

For the purposes of further comparison, these landscape typologies were collapsed into four more general groups, as shown in Table 9.1. All landscape categories that were based on tree cover were collapsed into the *junglat*/forest category. All agricultural categories were collapsed into a *kheti*/cultivated category. All fallow or grazing categories were collapsed into the category of *akad*/grassy. All categories signifying degraded, bare, or urban were collapsed together into *banjar*/bare. The images produced through this recategorization were then crosstabulated to assess the degree of coincidence (Table 9.3). The areal extent of each category is shown, as is the degree of intercategory agreement, expressed as a kappa value (following Eastman, Mckendry, & Fulk, 1991).

The spatial coverage of *kheti* land matches that of cultivated land extremely well, suggesting a high level of agreement between foresters and locals about what constitutes a field and where fields are located. There is slightly less overall agreement between producers' *akad* land and foresters' grassy/fallow land, though these two also seem to coincide overall. The long fallow and open

**TABLE 9.3. Producer Definitions for Areas
Identified as Forest by Professional Foresters**

Producer category	Coverage (in km^2)	Proportion (as %)
Junglat	263.27	59.28
Kheti	10.16	2.29
Gav/banjar	62.74	14.13
Akad/gocher	25.52	5.75
Not classified	82.44	18.56
Total	444.14	100.00

grazing lands of the region seem to have some categorical unity in the minds of both professionals and agropastoral locals. Agreement over what constitutes bare, degraded, or urban land is far poorer. There is extremely poor agreement between the *junglat* and forest categories. Producers recognize a larger coverage of degraded/bare land than do foresters, while foresters see considerably more forest than locals. It is evident that locals do not similarly name areas called forest by foresters. Breaking down the forester category coverage based on its definition by locals (Table 9.2), we see 263 square kilometers of forest cover, as defined by foresters, that locals would indeed call *junglat* or some comparable name suggesting or highlighting tree cover. Yet another 181 square kilometers, or more than 40% of the total, is differently identified. A small proportion of this area (around 8%) is identified as *akad* or *kheti* land, but a far larger area, almost one-third, is identified as either *banjar* (degraded or waste land) or simply left unclassified in local definitions.

Figure 9.2 shows the spatial distribution of agreement and disagreement in the forest category. The southeast portion of the image, where the wildlife sanctuary is located, is dominated by "shared" coverage that both foresters and locals would identify as tree-covered forest, or *junglat*. This is the old swath of deciduous forest that crowns the hilly areas running southwest–northeast across the region and which forms the geographical southernmost boundary of the Marwar region of Rajasthan. A unitary cultural and historical imaginary includes this kind of hilly tree cover as forest for both producers and foresters. The northwestern part of the image, representing the flat lowlands historically covered in savanna scrub, is dominated by land that foresters would define as forest but that local people would not. These areas producers prefer to call *banjar* (degraded) or to leave unclassified. But what is this cover?

Examining "forest" and "*junglat*" areas on the ground, it is evident that the variation in perception and definition reflects differences in tree species

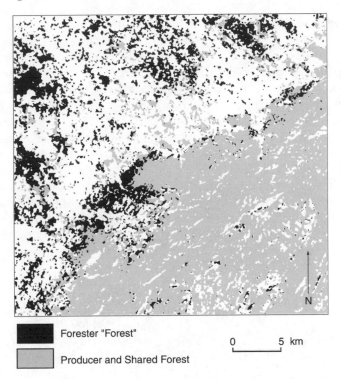

Forester "Forest"

Producer and Shared Forest

0   5 km

N

FIGURE 9.2. Comparing forest category coverage: the coverage of areas that both foresters and local producers define as "forest" includes the hilly areas of the reserve in the southeast; lowland land cover includes large areas that only foresters recognize as "forests."

cover at these locations. Those areas defined by foresters as "forest," which local producers do not so define, include areas where there has been a steady influx of the invasive species *Prosopis juliflora.*

Juliflora, or Mexican mesquite, is an aggressive and highly successful tree whose expansion has a great deal to do with its own inherent characteristics coupled with the increasingly vulnerable character of the regional landscape. *Juliflora* is drought-tolerant and has poor suitability for grazing or browsing by most animals. It is a leaf zerophyte that regulates transpiration losses and water shortages through leaf shedding. Its wood yield produces a remarkable 5–15 tons per hectare per year in plantation. The growth of its lateral root system improves soil conditions though soil fixation. The benefits of that soil improvement are not passed along to other species, however, since the leaves of the tree contain water-soluble allelopathic chemicals that halt germination of other species under its canopy (Noor, Salam, & Khan, 1995).

Land use dynamics also account for much of the species's success. Traditional village pastures and forests have been the sites of the most intense *juliflora* invasion. This is first because vast swaths of indigenous forest have been enclosed by the state for biodiversity preservation while traditional long-fallow pasturage has been decreased in the wake of state-sponsored irrigation subsidies. As displaced livestock are forced onto traditional commons, grazing, browsing, and coppicing pressures have increased. Second, however, by categorizing such traditional lands as "wasteland" and restructuring rights of use and access, the state revenue authorities have caused a collapse of traditional management authority. Together, this heavy pressure on the land has created a disturbance regime favored by *juliflora* (Robbins, 2001a, 2001b).

Even so, a central driving force behind the expansion of this landscape is the plantation of the species by foresters and the lack of institutional initiatives for its eradication, despite growing demands by local people. These are in turn tied to the varying perceptions of the species by competing groups. Statistical analysis of the data in Figures 9.2 and 9.3 reveals a relationship between forest-

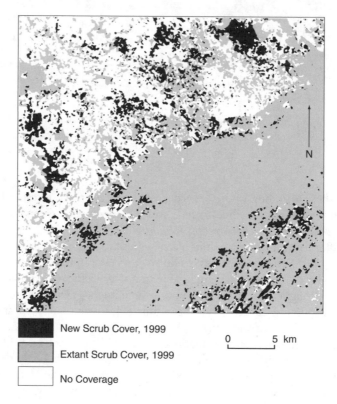

New Scrub Cover, 1999

Extant Scrub Cover, 1999

No Coverage

0   5 km

N

FIGURE 9.3. Emergent scrub canopy, 1986–1999: many parts of the region, especially those outside the forest reserve, are experiencing rapid afforestation

ers' "forest" and the invasive tree species; a significant proportion of the *juliflora* coverage are areas that foresters would call "forest" but local producers would not.

As suggested by the previous categorical analysis, local producers reject this coverage as forest, either giving it no name or describing it as "*banjar,*" or wasteland. As Pema Ram Divasi, an elder of the pastoral Raika community, stated emphatically when asked whether a thick stand of *Prosopis* trees represented a forest: "That is not a true forest. A forest has Dhav, Kair, and Palas. This has no name." For local people, the spread of this species is a problem owing to its low fodder value, inferior quality as a wood fuel, and concomitant contribution to grassland decline. They do not see the spread of *juliflora* as the spread of forest but, instead, often see it as a form, or at least a sign, of land degradation (Figure 9.4).

Most foresters, on the other hand, seem to recognize the cover as "forest" and encourage its growth, even though this new coverage of "forest" is expanding into grasslands far outside the range of the Forest Department enclosures, and even though it does not resemble indigenous forest coverage. Foresters act as a central vector for the spread of the tree through their continued plantation of the species in the region, despite an increasing agreement that the tree has little broader social value. Moreover, while there is increasing demand from lo-

FIGURE 9.4. *Prosopis juliflora* coverage in the village of Latara, 1990. This village pasture (*gocher*) was reportedly productive in the last 20 years but has since been overrun entirely by mesquite.

cal producers and even by many foresters for its eradication, the tree has not come under any form of state planning to remove it or even to halt its spread. This is centrally because the coverage of *juliflora* represents "forests" to forest-ers, who embrace the tree and its landscape, and support its growth, despite some misgivings, for profoundly structured reasons. The reasons are complex, but point to the relationship between forestry practice, remote sensing, and the bureaucratic incentives of institutionalized knowledge.

PORTABILITY, LEGITIMACY, AND OBJECTIVITY

It might be argued that the motivation for bureaucratic support of *juliflora* is based in the overall industrial orientation of Indian forestry. Certainly demo-cratic modernization of the Indian nation-state in the postindependence era enabled the growth of a powerful environmental bureaucracy that, like other bureaucratic structures, was created in the service of emergent capitalism and state industrialization (Corbridge & Jewitt, 1997; Jewitt, 1995; Gadgil & Guha, 1995).

Forestry in southern Rajasthan is notably noninstrumental, however, and reflects social welfare goals rather than cash-crop capitalization. *Juliflora* has shown little real commercial value, so that any simple explanation of bureau-cratic behavior in the support of *Angrezi babul* as an industrial product is hard to support. The Forest Department indeed extols the species largely for its power in "mitigating poverty" by supplying wood to avert a fuel crisis in the desert. In the case of Rajasthani land cover, a more robust explanation must go beyond simple instrumental logics and examine the social structure of envi-ronmental state agencies, where categorization, measurement, and perfor-mance evaluation together determine the tendencies and possibilities of eco-logical transformation (Dove, 1994, 1995). The politics of categorization are especially complex in the Indian bureaucracy, where "objective" standards are fundamental to the social structure of expertise.

Following Porter (1995), it is possible to see in the Kumbhalgarh case a struggle for objectivity, where the "private" knowledges of foresters is lever-aged into "public" knowledge by the establishment and promulgation of cate-gories that are remotely sensed, universal, and quantifiable.

Pursuing Objectivity through Mechanical and Quantitative Means

Measures, Kula (1986) noted, are an "attribute of authority," over which vying powers struggle for control and standardization. Similarly, the control of cate-gorical systems, like those for Indian land cover, represent a traditional area where state authorities seek the monopoly power to name. But more than this, scientific and bureaucratic communities like the Rajasthan Forest Department are impelled, by pressures imposed from without, toward specific practices, including pursuit of "mechanical" objectivity, quantification, and acceptance of universal analytical categories.

Mechanical objectivity represents replicable and rule-bound methods of practice, which becomes the normal practice when social and political pressures raise questions about the legitimacy of professional practitioners. Quantified and normalized data and proof reinforce the legitimacy of a bureaucracy that is otherwise suspected of politically instrumental decision making, and so establishes objectivity and trust in bureaucratic action. This pursuit of objectivity also hinges on the ability of differing communities (e.g., scientists, bureaucrats) to turn their "private" knowledges—those created and exercised locally (i.e., in the laboratory)—into "public" knowledge that can be deployed in a range of contexts.

Quantification and Species Choice

This drive for objectivity is reflected in Forest Department behavior in two ways. First, the system of incentives for promotion, and the hierarchical ordering of the forestry bureaucracy, is generally structured to reward advancement through rational and "objective" meritocratic methods; maintaining budgets, meeting schedules, and defending office autonomy are all rewarded with promotion. Forestry in Rajasthan, accordingly, evaluates and rewards its success largely based on objective indices of success: coverage figures and survival rates of trees (Robbins, 1998). The central statistical measures for plantation species selection are therefore population survival percentages and average tree heights; forestry means putting trees into the ground that will survive and grow tall. As a result, a range of exotic species were imported and evaluated over the last century, including *Eucalyptus* and a variety of *Acacia* species. Of these, *Prosopis juliflora* stands out repeatedly in the literature and in discussions with foresters as the tree of choice for arid areas.

As Hocking observes, "P. juliflora is a very aggressive tree, competes strongly for soil moisture and is difficult to eradicate once established" (1993, p. 281). In discussing the species with foresters it became clear that these characteristics were viewed as the very merits of the tree and that forester preference for the species was in part tied directly to the incentive structure of survival and advancement. As one forester reminded me, "You cannot be promoted if the trees do not survive." Thus, while the tree has its opponents within the bureaucracy, it is for the most part accepted without controversy. Its expansion and survival on the ground is a goal of forestry, and these trees therefore represent "forests."

Objectivity and Portable Technology

Second, the urge to mechanical objectivity is reinforced through the Forest Department's engagements with the diverse local populations with whom it is in contact and over whom it must attempt to enforce control. These populations, as demonstrated above, have their own secure and well-established set of "private" knowledges, the local categories of ground cover known to producers. The bureaucracy's legitimacy comes to be predicated, in reaction, on their ability to produce "public knowledge," the more universal, portable, measurable, and quantifiable land covers of state census. No matter how poorly such categories reflect the diversity of local ecologies, their establishment is necessary for state servants.

As a result, the rise to prominence of *juliflora* is tied to the emergence of remote sensing from aerial photography and satellite images platforms, especially in state-sponsored projects that form an increasing part of India's social forestry mandate. Budget allotment for forestry activities in Rajasthan between 1994 and 1997 expanded by 255%, increasing by 15 million $U.S., while the total area under Forest Department control expanded by only 3.6%, with the net acquisition of around 1,100 square kilometers of land. These budget increases have therefore served less to enclose land than to increase plantation and expand and professionalize the bureaucracy, train staff, and implement new technologies.

In particular, air photography and remote sensing have long been targeted for incorporation into management and are increasingly used to map areas of forest cover and track progress in afforestation and reforestation. In the area of "wastelands development," there has also been a growth in inventory and mapping activity. Between 1985 and 1990, wastelands development in Rajasthan received 390 million rupees (around 10 million $U.S.) from the United States Agency for International Development (USAID) and the World Bank. Much of this funding went directly into plantation efforts, bringing 120,000 hectares in the state under direct management, but like traditional forestry efforts, these projects were predicated on mapping efforts conducted in collaboration with the National Remote Sensing Agency and the Survey of India. These endeavor to map "wastelands" and identify areas for activity and progress made by the Department of Forestry and other agencies within the Ministry of Environment.

Though few of these tools are yet available to the local forest managers who actually oversee forestry plantation and administration, the overall awareness of the implications of these forms of spatial data is commonly understood throughout the bureaucracy. The connection between development, mapping, and remote sensing is an increasing part of the daily vocabulary of forestry. Specifically, foresters generally report that the rapid spread of the

Angrezi babul (*Prosopis juliflora*) tree can be seen from the air, and that this transformation stands as a testimony to the success of the Forest Department's public mission (Robbins, 2000). This point of view, which champions the species by way of its *observable* canopy, was consistently reiterated in interviews with forest guards, range officers, and more elite managers.

Thus, as foresters adopt remote sensing techniques or at least become increasingly aware of them, the bureaucratic measurement of success—coverage of land in something called "forest"—increasingly demands an understanding of "forest" as a dense lateral green canopy that is spectrally discernable, or at least that can easily be observed from the air. Such a definition matches the characteristics of the monocultural stands of *juliflora*. Having defined this form of canopy cover as "forest," foresters largely ignore or tolerate it, or, in any event, do not define it as a problem. These technodiscursive feedbacks in the establishment of *juliflora*-dominated landscapes are summarized in Figure 9.5, which shows the relationship between the technologies of remote sensing, the ecological tendencies of *Prosopis juliflora*, and the discursive construction of "forests" in the landscape. Introduced technology creates the parameters within which bureaucratic incentives are formed; *juliflora* is overlooked as benign since the greening of the landscape, visible with the introduction of new technology, matches the incentives of the agency. The technological and categorical structure of bureaucratic forestry sets and reinforces the direction of environmental change.

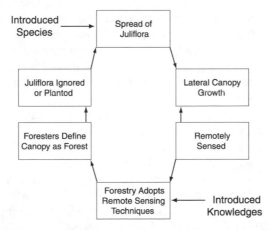

FIGURE 9.5. Technodiscursive feedbacks: the introduction of new species and new systems of environmental knowledge propels the ongoing and increased acceptance/tolerance of *juliflora* land cover, which in turn reinforces diffusion of the tree itself.

Technological Agency and Landscape

Ultimately, then, satellite images act in at least two ways not otherwise noted in celebratory discussions of the power of the tool. First, following Latour (1987), the satellite appears as the arbiter for a dispute about land cover but, in fact, actually acts to justify an already settled dispute about the nature of that landscape. Second, following Veregin (1995) and Winner (1977), the satellite image serves as a force for reverse adaptation, where existing landscapes are reengineered to suit technical means, rather than the other way around.

In the first case, the satellite image serves a complex double role in the interpretation of landscapes. The image appears as a "natural" artifact of reflectance that settles the question "How much forest is there?" But, as we have seen, the image is created only after a prior argument is settled: "What are forests?" Once we have decided that mesquite monoculture constitutes a forest landscape, then controlled and statistically reliable methods of generalization from multiband data shows that a wide and expanding swath of forest exists on the plains. Thus satellite images are not only the cause of the settlement of a scientific controversy ("How much forest is there?") but are in fact themselves the consequence of a dispute settlement ("What are forests") (Latour, 1987). By appearing as the natural arbiter of the former dispute, the latter one disappears altogether and landscapes are written as though there was no disagreement over the nature of the land. The satellite image ultimately enforces an interpretation of the landscape rather than arbitrating between competing claims.

In the second case, satellite and image-processing GIS technology have inherent tendencies for interpreting and defining space, and so act to alter the landscape in particular ways. In this case, the monocultural stands of *juliflora* form coherent pixels and easily reveal themselves to remotely sensed technologies. As a result (and owing to its overall persistence and survival), the species comes to represent "forests" for foresters, who then act or fail to act such that such forest covers spread.

The technology, therefore, is not neutral. Its optics cause it to participate in landscape change in specific ways. Landscapes are adapted to suit the observational parameters of the tool. This is a "reverse adaptation" of geography by geospatial technology (Winner, 1977; Veregin, 1995).

A "normal" or intuitive model of technological development holds that the parameters of a problem or system are recorded and encoded in technology to help describe or manage that system. Consider accounting, for example, a technology designed to describe business processes and the flow of capital. But as the technology increasingly embeds itself into the workings of the system, it comes to alter the system through its own metrics. Business practices are increasingly adapted so they will appear in a specific form on an accounting sheet! The technology has reverse-engineered the system, with the participation of many actors, including accountants and shareholders seeking to take home the largest dividends.

The case of *juliflora* is perhaps more profound (or disturbing) since the adjusted system is one we tend to think of as "natural" and free, if not from human influences, at least from the influences of the distant devices we use to measure them. So too the actors in this case include foresters, grazing animals, and trees, as well as satellites and air photography flights. The satellite image is therefore a social agent of environmental change, serving as an ally in disputes over the nature of the landscape while actually acting to help create new incentives that remake the land in its own form of measure.

The case of Kumbhalgarh, moreover, is in no way unique. The definition of "forest" land cover by Indian bureaucrats is very much like other cases around the world, including the definition of "old-growth forest" in the U.S. Pacific Northwest or the establishment of ecosystem boundaries in Toronto suburbs. In a comparable case from U.S. forestry, chemical inputs of fertilizer and pesticides increased as a result of changes in accounting techniques, which were themselves obligated by congressional mandates for conservation (Porter, 1995).

In sum, I have argued that the arrival of remotely sensed data, while extremely promising for inventory and analysis in complex landscapes, brings its own limitations and biases emphasizing horizontality and reflectivity in landscapes. Coupled with the bureaucratic incentives of self-reproduction, these tendencies are set loose to reproduce landscapes in their own form, green canopies with dubious human or ecosystem value. The measurement of these resulting landscapes through the very tools of their transformation naturalizes the resulting ecologies and erases the history of intervention from which they arise. Only by backtracking to the production of such images, and showing the contentiousness of the categories that undergird the deployment of imagery, can the effects of landscape analysis be rendered clear.

I have not argued here that satellite imagery and air photography are bad tools for the analysis and management of natural systems. These instruments continue to be some of the most promising for exploring landscape change, and assessing human–environment action across scale (Liverman, Moran, Rindfuss, & Stern, 1998). Nor have I suggested that the Indian forestry bureaucracy, charged as it is with a staggeringly difficult social–environmental mission, is a malicious agency or inherently ill-suited to land management. The state will continue to be a major player in south Asian social ecology, for better and for worse, and the historic lack of engagement with state agencies on the part of Indian environmentalists is likely to be self-defeating (Rangan, 1997). Rather, I have argued that the encounter of remote sensing with bureaucratic authority under conditions of ecological modernization is one of which analysts and activists must be wary.

As a policy issue, this provides a methodological opening for activist possibilities and progressive alternatives, but it also provides a cautionary conclusion. Satellite images are only meaningless reflectance values: they can be interpreted through highly reliable statistical procedures and GIS analyses using the categorical reality worlds of *anyone*, from foresters and land revenue officers to herders and farmers. There is no essential linkage between the numbers and our understanding of the landscape from which they are derived. The techniques described here can be reproduced in the planning process as a form of "countermapping" (Peluso, 1995), where the visions of the landscape from place-specific understandings may be generalized to powerful map surfaces.

As a lesson of caution, however, it suggests that those researchers and activists interested in democratic and sustainable planning must place the apparent objectivity of satellite data and air photography under careful scrutiny; they only open, and cannot close, debate on the state of the world. Their use tends to direct attention toward certain features of the environment while disguising others, while their presence tends to depoliticize what is an inherently political process: the interpretation of landscape. More than this, however, it suggests a need for special caution in approaching environmental bureaucracies infused with, and adapted to, this technology. The tail can too easily wag the dog where geospatial tools are concerned.

While we will inevitably and increasingly depend on remotely sensed data to measure the nature and direction of ecological change, we can not depend on such imagery to adjudicate fundamental disputes over the nature of the environment that are prerequisite to such measurements. And though we must use the tool to measure the rate and extent of environmental change, we must acknowledge that the employment of the satellite itself serves as an agent of environmental transformation. Living with such contradictions is a prerequisite to future geographic science.

ACKNOWLEDGMENTS

This work was made possible with support of the American Institute of Indian Studies and Ohio State University. Special thanks go to Hanwant Singh Rathore at the Lok Hit Pashu Palak Sansthan, Ilse Köhler-Rollefson at the League for Pastoral Peoples, and S. M. Mohnot at the School for Desert Sciences. Thanks also to the foresters and residents of Sadri and Mandigarh and to Anoop Banarjee and Sakka Ram Divasi.

NOTES

1. This analysis follows on the method described and preliminarily deployed in Robbins (2001b; Robbins & Maddock, 2001), which describes the epistemological underpinnings of a "participatory classification technique."
2. For fuller discussion of the methodology, see Robbins (2001a).

REFERENCES

Barnes, T. J., & Duncan, J. S. (1992). Introduction: Writing Worlds. In T. J. Barnes & J. S. Duncan, *Writing worlds: Discourse, text, and metaphor in the representation of landscape* (pp. 1–17). New York: Routledge.

Corbridge, S., & Jewitt, S. (1997). From forest struggles to forest citizens? Joint forest management in the unquiet woods of India's Jarkhand *Environment and Planning A, 29,* 2145–2164.

Dove, M. (1994). The existential status of the Pakistani farmer: Studying official constructions of social reality. *Ethnology, 33*(4), 331–351.

Dove, M. (1995). The theory of social forestry intervention: The state of the art in Asia *Agroforestry Systems, 30,* 315–340.

Eastman, R., Mckendry, J., & Fulk, M. (1991). *Change and time series analysis.* Geneva: UNITAR.

Ellen, R. (1982). *Environment, subsistence and system: The ecology of small scale social formations.* Cambridge, UK: Cambridge University Press.

Gadgil, M., & Guha, R. (1995). *Ecology and equity: The use and abuse of nature in contemporary India.* London and New York: Routledge.

Hocking, D. (Ed.). (1993). *Trees for drylands.* New Delhi: Oxford University Press/IBH Publishing.

Jewitt, S. (1995). Europe's "Others"?: Forestry policy and practices in colonial and postcolonial India. *Environment and Planning D: Society and Space, 13,* 67–90.

Kula, W. (1986). *Measures and men.* Princeton,NJ: Princeton University Press.

Latour, B. (1987). *Science in action: How to follow scientists and engineers through society.* Cambridge, MA: Harvard University Press.

Liverman, D., Moran, E., Rindfuss, R., & Stern, P. C. (Eds.). (1998). *People and pixels: Linking remote sensing and social science.* Washington, DC: National Academy Press.

Noor, M., Salam, U., & Khan, M. A. (1995). Allelopathic effects of Prosopis juliflora swartz. *Journal of Arid Environments, 31*(1), 83–90.

Peluso, N. (1995). Whose woods are these? Counter-mapping Forest Territories in Kalimantan, Indonesia. *Antipode, 27*(4), 383–388.

Porter, T. M. (1995). *Trust in numbers: The pursuit of objectivity in science and public life.* Princeton, NJ: Princeton University Press.

Rajputana-Gazetteers. (1908). *The western Rajputana states residency and Bikaner.* Gurgaor, India: Vintage Books.

Rangan, H. (1997). Indian environmentalism and the question of the state: Problems and prospects for sustainable development. *Environment and Planning A, 29,* 2129–2143.

Robbins, P. (1998). Paper forests: Imagining and deploying exogenous ecologies in arid India. *Geoforum, 29*(1), 69–86.

Robbins, P. (2000). The practical politics of knowing: State environmental knowledge and local political economy. *Economic Geography, 76*(2), 126–144.

Robbins, P. (2001a). Tracking invasive land covers in India; or, Why our landscapes have never been modern. *Annals of the Association of American Geographers, 91*(4), 637–659.

Robbins, P. (2001b). Fixed categories in a portable landscape: The causes and consequences of land cover classification. *Environment and Planning A, 33*(1), 161–179.

Robbins, P., & Maddock, T. (2001). Interrogating land cover categories: Metaphor and method in remote sensing. *Cartography and Geographic Information Science, 27*(4), 295–309.

Veregin, H. (1995). Computer innovation and adoption in geography: A critique of conventional technological models. In J. Pickles (Ed.), *Ground truth: The social implications of geographic information systems* (pp. 88–112). New York: Guilford Press.

Weiner, D., Warner, T. A., Harris, T. M., & Levin, R. M. (1995). Apartheid representations in a digital landscape: GIS, remote sensing and local knowledge in Kiepersol, South Africa. *Cartography and Geographic Information Systems, 22*(1), 30–44.

Winner, L. (1977). *Autonomous technology: Technics-out-of-control as a theme in political thought.* Cambridge, MA: MIT Press.

CHAPTER 10

GIS Representations of Nature, Political Ecology, and the Study of Land Use and Land Cover Change in South Africa

Brent McCusker
Daniel Weiner

Recent "GIS and Society" literature and debates have focused on social, spatial, and political issues with little concern about nature and the environment. This is curious given the prominence of environmental information in populating GIS (geographic information system) databases. GIS has its roots in land information systems and produce representations of nature that shape how environmental resources are perceived, controlled, and exploited.

This chapter links geospatial technologies and political ecology in two ways. First, we examine how nature is represented in a scale-dependant GIS analysis of land cover and land use change in southern Africa. We argue that the use of geospatial technologies to study landscape change has led to differing, and sometimes contradictory, representations of nature. As a result, how nature–society relationships are conceptualized is impacted by the technologies employed and what appear as "natural" landscapes are historically and politically constituted in a digital environment. A South Africa land reform case study then provides an example of how political ecological processes might be incorporated within a GIS. This is done through participatory fieldwork on recently redistributed farms where GIS-based spatial analyses of land use and land cover change are compared and contrasted with local narratives

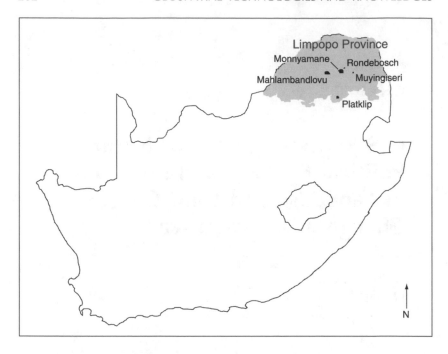

FIGURE 10.1. Location of study sites in South Africa.

that were derived from intensive interviews, transects walks, and community ground truthing (Figure 10.1).

The case study demonstrates the potential for merging political ecology and GIS; hidden political ecologies are uncovered and the GIS analysis became more sensitive to local context. In the chapter, we define GIS broadly to include a range of geospatial technologies, including remote sensing and geographic positioning systems (GPS).

GIS, NATURE, AND SOCIETY

There is a now a substantial critical social theory literature concerned with the social, political, and epistemological impacts of the diffusion of geographic information systems in society (see, e.g., Aitken & Michel, 1995; Curry, 1994; Goss, 1995; Harris & Weiner, 1998; Lake, 1993; Oppenshaw, 1991; Pickles, 1991, 1995; Sheppard & Poiker, 1995; Sheppard, Couclelis, Graham, Harrington, & Onsrud, 1999; Taylor & Overton, 1991). There is also a rapidly growing interest in the prospects for participatory GIS, remote sensing, and GPS applications (Craig, Harris, & Weiner, 2002; Kwaku-Kyem, 1999; Obermeyer, 1998;

Rundstrom, 1995). These two very important discussions have in common the privileging of sociopolitical issues with little thought about the intersection of GIS, the environment, and society.

An exception is *People and Pixels* (Liverman, Moran, Rindfuss, & Stern, 1998), an edited volume that attempts to fill a "gap" between remote sensing applications and social science by linking the social (people) with the technical (pixels). The goal of the text is to augment the analysis of social processes with remotely sensed images. For example, Geoghegan et al. explore Markov modeling to predict land use and land cover change in eastern Maryland, while Wood and Skole link satellite imagery and census data in the Brazilian Amazon. Moran and Brondizio examine land use change after deforestation in Amazonia, and several other contributors detail similar applications that attempt to incorporate socioeconomic data with remotely sensed images. These contributions demonstrate the utility of geospatial techniques for merging social science and remote sensing and identify important methodological issues. For example, Rindfuss and Stern demonstrate that remote sensing can "sometimes provide time-series data of good comparability . . . on variable of interest to social scientists concerned with the effects of context on behavior or with processes of human–environment interaction" (cited in Liverman et al., 1998, p. 9). Geoghegan et al. note that "creative explanatory variables can be constructed from remotely sensed data through the use of GIS, as with landscape patterns or land-use mosaics" (cited in Liverman et al., 1998, p. 57).

A concern with *People and Pixels* is the type of social science that is being integrated. There is an overreliance on economic indicators of change merged with satellite imagery, leading to a narrow reading of critical social and technical issues. The type of integration pursued fits nicely into the modeling paradigm of global change studies, but fails to seriously engage with political ecology and critical social theory more generally. This is the objective of a recent paper by Turner (2001). Working in the Sahel region of Africa, he argues that

> the increased use of remotely-sensed data and GIS techniques has not revolutionized the ways in which we analyze people–environment relations in the Sahel but actually further entrenched traditional modes of analysis. These modes of analysis rely heavily on visual descriptions of land cover as measures of environmental change and evaluate causes of environmental change through spatial correlation with little understanding of local context. Interestingly, it is these forms of environmental analysis that new understandings of dryland ecology have seriously questioned. (p. 10)

This is an important argument and mirrors the "GIS and Society" concerns about the impact of geospatial technologies on geographic methods, philosophies of science, and ways of knowing. Turner (2001, p. 13), however, also argues that "cultural and political ecologists can gain from the incorporation of these technologies in their work but in so doing they need to be critically self-reflective about the data demands of these technologies and how such data demands may subvert their broader research goals."

The goal of Harris, Weiner, Warner, and Levin (1995) was to incorporate socially differentiated cognitive maps to research regional political ecologies in South Africa's Mpumalanga Province (see also Weiner & Harris, 2003). They found that local knowledge could easily be incorporated into a multimedia GIS and was a useful addition to existing "expert" data provided by government agencies. Specifically, they were able to demonstrate how GIS could be a technological platform for studying the historical geography of forced removals, perceptions on land potential and local land uses, and possible locations for land reform projects. The participatory GIS incorporated landscape power relations and political struggles and provided some insight into how political ecology might be embedded within a geographic information system.

The few case studies that have integrated geospatial technologies with political ecology indicate significant potential for conceptual and empirical research. For example, GIS is ideal for identifying local-level environmental variations and spatial and historical relationships that help to locate *ecology* in political ecology. GPS can be ideal for identifying and mapping community spatial stories, while remotely sensed images provide vivid visualizations of socially produced landscapes and how communities adapt to microenvironments and environmental change. These technologies are thus well suited for analyzing natural resource access, for examining use and ownership, and for mapping landscape power relations and political struggles. In particular, multimedia Internet GIS brings together valuable qualitative and quantitative information for potentially robust and placed-based political ecological research. This integration of political ecology and geospatial technologies is, however, scale-dependent.

POLITICAL ECOLOGY AND SCALE

The social sciences, and geography in particular, have struggled with the issue of scale. In land use and land cover change studies, scale has often been examined as an independent variable, as a container through which processes flow. We reject the notion that the goal of generalization in human–nature science is simply to "fill in" the middle scales (regional, national) in an ongoing search to link global observations to local processes (Gallopin, 1991). This viewpoint has its origins in the very foundations of GIS and GIScience, where data are aggregated and disaggregated in order to fulfill the requirements of modeling. In fact, the overemphasis on modeling land use and land cover change has sidetracked the discussion on linking scales of analysis. Models are used to constitute information about regions instead of directly examining regions themselves. In this respect, we set aside attempts by modelers to link scales; we seek to uncover processes that are made manifest across and within scales, not to discuss derived estimates of them.

Scale in the study of land use and land cover change is a salient yet unresolved issue (Campbell, 1998). A political ecology approach could be useful if strong relationships between global and local processes were readily made manifest. One of the central difficulties in addressing scale in land use and land cover change analysis is the use of data from different scales. Political ecologists tend either to project empirical evidence directly from local-scale case studies or to focus on regional and global impacts on local landscapes.

To date, a *dialectical* political ecology (Blaikie, 1994), one in which an attempt is made to examine the relationship between the scale of analysis and the manifestations of political ecology, has not yet materialized. One possible way to do this is through the integration of GIS in nature–society studies. In global change studies, proximate variables, such as population, have been used to attempt to theorize and model land use change at the regional and global scales. However, the data present a problem. Meyer and Turner (1994) argue:

> Though a global view is required for some purposes, a globally aggregate one is insufficient for answering many pressing questions. The net worldwide trajectories of land-cover change are rarely duplicated in any region or locality. Consequently, explanations, forecasts, and prescriptions developed only from global aggregate data are likely to be worse than useless when applied in subglobal units. Nor can adequate global projections be developed from global aggregate data alone because global tools represent aggregations of quite dissimilar world regions. (p. 7)

While discounting the global approach due to a lack of reproducibility in regional and local scales, the authors point out the problems with the local- and regional scale approach as it relates to land use and land cover change studies:

> Yet the opposite extreme from the global aggregate approach, a plethora of microstudies highly attentive to local context and singularity, is equally unsatisfactory given the needs and constraints of the global change research program. A large literature of small-area studies does exist, and it offers many insights into the complexities of nature–society interactions. Practical considerations, though, prohibit the separate study for global modeling purposes of every piece in the world's mosaic of environmental and socioeconomic conditions. Nor could the results, even if collected in a systematic and comparable way, necessarily be aggregated unproblematically for higher scale of analysis. (p. 7)

Lambin (1992), describing desertification in Burkina Faso, remarks that "an analysis of the environmental consequences of decision-making often requires a broadening of geographical scale" (p. 5). The argument here is that upward linkages in scale cannot be overlooked when conducting specific case studies. Further, he writes, "An approach that employs a nested set of spatial scales has proven to be appropriate to understand the behavior of land managers responsible for desertification" (p. 4). In reality, many studies tend to focus on one scale of analysis with the others acting as inputs and outputs. In what he refers to as "scaling parsimony," Turner (2001) rejects the idea that "more parsimonious explanations exist when proposed causal factors work at the same spatial scale" (p. 191) and employs the notion that "an understanding of the dynamics of regional land use change requires moving from correlations at socially abstracted spatial scales toward political–ecological studies that not only focus on explanations for local land use changes but consider the aggregate effect of these changes at the level of the region" (p. 192).

In the remainder of this chapter, we explore how GIS and remote sensing technologies have impacted representations of nature and society at a regional scale and then examine the prospects for scaling up with a GIS-based local political ecology analysis of South African land reform.

REPRESENTATIONS OF
THE SOUTHERN AFRICAN LANDSCAPE

Southern Africa is a region with diverse sociopolitical and environmental histories. At a local scale, these landscapes are visible as degraded migrant labor reserves, intensive and underutilized white settler estates, African township and periurban settlements, ecotourism parks, and so on (Harris et al., 1995). These socially produced landscapes are, of course, mediated by local environmental context, which is also quite variable over space. Plate 10.1 displays a 1992–1993 regional map of southern African land cover produced from advanced very-high-resolution radiometer (AVHRR) data at 1-kilometer resolution. This representation of the region naturalizes the landscape and eliminates the human layer of this region's nature–society interaction. Plate 10.2 demonstrates that the resolution of the data can mask sociopolitical landscapes. The top plate is a subset of the continental-scale image derived from the AVHRR data. The center plate shows the same area for approximately the same time period with a very visible divide between the former black homelands on the left half and the former white-occupied Transvaal on the right. This visible environmental difference is eliminated by the scale of mapping, thus hiding important nature–society relationships. While the border is clearly evident in the Landsat (satellite) image, in the AVHRR image the same stark environmental transition is very difficult to discern. Referenced by technicians with more knowledge of the local environment, the regional interpretation from the Satellite Applications Centre/Centre for Scientific and Industrial Research (SAC-CSIR) of South Africa displays the border. This representation shows that while it is a relict boundary, it still differentiates land use and land cover patterns in the landscape. However, the differentiation is not as clear as when observing the raw Landsat imagery. Without careful analysis, a major feature of the social and environmental landscape can be reduced in significance, if not hidden altogether.

Global and subglobal scales of analysis, especially in land use and land cover change, have also been constrained by data issues. While finer resolution imagery has recently been made available (e.g., Ikonos satellite imagery), several issues constrain global mapping and analysis. First, data management concerns would immediately overwhelm the analysis. There would simply be too much data to handle. Second, interpretation of the imagery would be very time-consuming, either through automated or manual techniques. Third, cost considerations would prevent such fine-scale global- or subglobal-scale mapping. The high scene price of fine resolution data and the number of images needed prohibits this type of analysis. Fourth, and problematic at all scales, is the temporal inconsistency of analysis. Choice of year or season of analysis is often determined not by important socioeconomic or environmental events, but rather by the degree of cloud cover in a particular image. As a response, mapping and monitoring of land cover and land use has occurred at the global and subglobal scale at very coarse resolution (such as 1 kilometer). While this provides an *indication* of land cover, it remains inadequate for the political ecologist who seeks to link micro-, meso-, and macroscales and to represent how power and politics at the landscape level impact the way in which environments are represented and exploited.

INTEGRATING POLITICAL ECOLOGY AND GIS IN LIMPOPO PROVINCE

Many nations, especially in the developed world, have undertaken national land cover mapping. South Africa is one of the few developing nations to have undertaken a detailed inventory of its land cover (1994–1995). As such, this provides at least some basis for comparison between scales. The land cover map of South Africa was generated through labor-intensive manual interpretation, a difficult and costly method to reproduce (Plate 10.1).

Our integration of geospatial technologies with political ecology includes a comparison of top-down provincial land cover mapping for Limpopo Province with bottom-up mapping at a local scale. A bottom-up map was produced with data from five land reform projects and the adjacent community in the province and linked community input with data derived with geospatial technologies (McCusker, 2001). In this chapter, we only show the results from one project as an example (Figure 10.2). This local-scale integration produces a

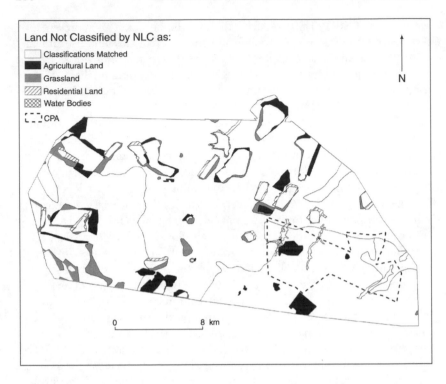

FIGURE 10.2. Divergence in land cover classification between national land cover dataset and McCusker (2001).

political ecology that combines "expert" data with local knowledge in an attempt to understand patterns at the provincial scale.

An overlay analysis shows this nationally derived provincial map and a locally interpreted one. Large areas of similarity and divergence are evident. The major differences between the maps include misclassifications inherent when using coarse-resolution imagery. The largest areas of difference are on the edges of agricultural land, in agricultural land where fallow was occurring, and where the method of planting of crops led to a spatial pattern that fell below the minimum mapping unit. Additional causes of divergence include variations in interpretations of land cover by technicians, differences in the imagery used for interpretation, and incorporation of community local knowledge in image interpretation. What the provincial map does indicate is that as one moves closer to the local scale, assessments become more refined and divergence with the locally derived maps lessens.

Given that the exact same images were not used, neither spatially nor temporally, variance can arise. The areas under examination include places where the African practice of shifting cultivation still occurs. Thus, an image obtained for a spot in 1994 will represent a different landscape from one used just 1 year later. This is a problem in the use of remotely sensed images, one that has not been adequately addressed.

Typically, the technical community has addressed time as a linear element. Thus using imagery from different years for interpretation has not been widely identified as a problem, or where it has been so identified, alternative solutions have been pursued. But time is not always linear, nor does it operate according to the amount of cloud cover in a scene. Analyses of large regions' land cover change routinely use temporally divergent images to compare spatially adjacent areas (e.g., scene left will be from 1992 while scene right will be from 1993). Not only does the landscape become temporally jaded, but the political ecology of the landscape can be contradictory. In the final interpretation, "time" in scene left and "time + 1" in scene right is transformed into one temporal representation. Thus spatially explicit projections of processes in the landscape are socially and temporally disjointed, at best. This becomes even more problematic when scaling up. Landscapes become homogenized not only in space, but also through time. Thus, aggregated analyses of land cover represent *multiple, overlapping, and contrasting political ecologies* within the same GIS coverage. It is no wonder, then, that many geographers have struggled so greatly with the regional scale of analysis. The third explanation for the divergence between the scenes lies in the process behind which the local representation was created. The local image relied heavily on the incorporation of participatory ground truthing, which we turn to next.

Land Reform and Changes in Land Cover and Land Use

The study of local landscapes is where GIS and political ecology are most compatible. Since the democratic elections in 1994, the South African government has initiated a land reform program. The broad objective of this initiative is to redress South Africa's historical political ecology, which features gross inequities in land and water access (Levin & Weiner, 1997). Specifically, South African land reform is intended to

> deal effectively with: 1) the injustices of racially based land dispossession of the past; 2) the need for a more equitable distribution of land ownership; 3) the need for land reform to reduce poverty and contribute to economic growth; 4) security of tenure for all; and 5) a system of land management which will support sustainable land use patterns and rapid land release for development. (Department of Land Affairs, 1997, p. 7)

One important feature of this program is the creation of Community Property Associations (CPAs; see Figure 10.1). These are community management structures created for the ownership and administration of land reform projects, but they do not necessarily result in collective production. An assessment of the CPA program was conducted in 1999–2000 in Limpopo Province to study changes in land cover and land use associated with the new land reform program. Five land reform projects were investigated; two are presented here as representative.

The Mahlambandlovu CPA is located approximately 20 kilometers from Pietersburg just across the border of the former Lebowa homeland in the highveld. The farms that constitute the CPA were purchased from a (white) large-scale commercial farmer in 1997 and formally established as a full-fledged CPA in 1998. The CPA has a small truck, two tractors, and an array of farming implements with which they raise chickens and cattle. The farmland for the CPA is located in a lower potential zone for farming, receiving lower rainfall (400–600 millimeters/year) and with generally less productive soils. CPA membership stands at 396, but the secretary estimates that there are fewer than 160 active members. After 3 months of participatory fieldwork, 120 members were interviewed.

The Rondebosch CPA is situated in the northern Drakensburg Escarpment. The 30 members of the CPA reside in the Sekgopo community, which lies just to the northeast of the actual CPA land. The farmland of the CPA is in a higher potential area that receives more rainfall (600–800 millimeters/year) and has more productive soils, especially when compared to the Mahlambandlovu CPA. The Great Letaba River separates the community from the farm. In late 1999 and early 2000 devastating floods struck northern South Africa, including the Sekgopo community and the Rondebosch CPA. Widespread damage was observed at the farm and the community, including the destruction of most of the road network and the CPA's water pump, crops, and livestock.

To analyze land use and land cover change associated with these land reform projects, we employed a range of GIScience methods. Six Landsat Thematic Mapper images were obtained for two time periods: 1988 and 2000. The analysis also included a 1995 reference scene. Aerial photographs were also obtained, but for a longer time period (1970–1994). The aerial photographs were used for reference only, as they were neither spatially or temporally complete or consistent. The images were then processed and interpreted, and selected areas were digitized. This provided a series of maps for the transferred land of the CPAs. These maps were then interpreted for land cover, land use, and change. Community participants were invited to discuss their ideas and perceptions of land use and land cover change through a quantitative survey of 400 participants, group project meetings, informal discussions in the field, and intensive individual interviews. The analysis also included individual and group transect walks with GPS receivers.

Land cover maps were produced from detailed interpretation of the images and through equally detailed fieldwork. Participants were asked to describe the land cover in terms to coincide with the years of the satellite images. The respondent would stand and point to specific plots of land and describe the land cover in the late 1980s (1988) and then describe changes that had occurred in the intervening period. The GIS maps were produced before entering the field, so there was already an indication of land cover and land use change. Because of this very explicit analysis, land cover and land change issues could be explored on specific plots and compared with the GIS maps. Respondents were given detailed explanations of the GIS-derived maps and asked to describe why changes happened in certain areas. The maps were sufficiently detailed for respondents to explain specific land uses and reasons for change. The large sample size of the quantitative survey allowed for verification and comparison of perceptions of change.

The study of land use and land cover change included an analysis of production intensification and extensification, where *intensification* refers to the shift in land use from a state where value of product per land unit increases, while *extensification* refers to shifts in land use where the value of product per land unit decreases. In the context of this study, a shift from grassland or grazing land use to an agriculture use would be considered intensification, while the reverse would be considered extensification. Although not agricultural in nature, a shift from either grassland/grazing or agricultural use into residential use would also be considered intensification, since at least theoretically the value of the land per unit would increase with structures being built on it, even when in reality this is not always the case.

On the Rondebosch and Mahlambandlovu CPAs extensification and abandonment occurred. Areas that had been more intensively utilized in 1988 were less used in 2000. Areas outside of the actual redistributed farms, however, showed a variety of land use changes. This "external" change is mentioned here, as the owners of the farms do not actually occupy the transferred lands. In the following descriptions of land use change on individual CPAs the term "agriculture" refers to both farming and herding, while "grassland" refers to spatially undifferentiated grassy scrubland. "Grassland" should not necessarily be taken to mean land on which cattle are raised. Table 10.1 quantifies land conversion for the study areas. In the areas where the most intensification was anticipated, those farms transferred under the CPA program, extensification was observed.

The Mahlambandlovu CPA is the largest of the CPAs in the area and the largest in terms of membership in the study group. As evidenced by Plate 10.3, the membership is spread across a large rural area to the west of the CPA. On the CPA itself, land use change consisted largely of extensification. In the western area of the CPA (letter A) land that had been used in 1988 as farmland has since become grassland. Outside of the CPA, where the membership resides scattered among other nonmembers, shifts from grassland to

TABLE 10.1. Area Change by Land Use: 1988–2000

Land use and cover	CPA change	Change on areas outside CPA (members' homes)
Agriculture	–20.0%	–4.0%
Grassland	+14.9%	–20.1%
Wooded	–6.2%	–0.4%
Residential	+2.7%	+22.4%
Water bodies	+9.6%	+2.1%

Note. k-hat for error matrix = 89.3%.

agriculture are evidenced just to the north of the CPA (letter B), while a reverse shift can be seen to the south and southwest of the CPA. Noticeable expansion of settlements (rural townships) is shown at letter C on Plate 10.3. The overall pattern of change on the CPA itself, then, has been extensification.

The area around the Rondebosch CPA (Plate 10.4) has experienced a large growth in population since 1988. The area to the north of the CPA (letter D) shows a dramatic increase in the amount of land area used for settlement. In addition, changes have occurred since 1988 in areas other than the actual redistributed farm. On the CPA farmland itself, extensification has occurred. Areas that were clearly used for agriculture in 1988 have reverted to thinly wooded areas (letter E). The membership of the CPA consistently remarked about the lack of equipment, labor, and skills necessary to efficiently utilize the CPA farmland. The mean age of the respondents on the CPA was 50–59. An older population would certainly find it difficult to farm the land without the assistance of younger people.

GIS and Hidden Political Ecologies

Several examples from the above exercise yield useful insights for merging geospatial technologies and political ecology. Figure 10.2 shows the areas of divergence between a nationally derived land cover assessment and the authors' assessment that incorporated local political ecology. A similar divergence between representations occurred when the authors were creating the original local land cover maps. Analysis of the imagery was conducted in the United States and then transported to the field. Two simultaneous processes were occurring during fieldwork. The first was correcting technical errors, or ground truthing. The second was more contextual: identifying and investigating local environmental and political issues behind the images. During this exercise two patterns clearly emerged. First, there were areas on the imagery where an understanding of the local political ecology helped in interpretation.

Some areas were unclear on the satellite imagery or assumptions were made that fell away when local knowledge was incorporated into the interpretation. Across all images in the study sites, this accounted for a great deal of the reinterpretation. However, a second pattern also emerged, one that was less easy to explain. In some images, there were clear shifts from one land cover type to another where local knowledge did not concur.

One such example was identified near the Mahalambandlovu CPA. During a transect walk, a community member noted that the author (BM) had classified a certain plot of land as grassland in the 1988 imagery and as agricultural land in the 2000 imagery. He suggested that there must be an error because the community had used that plot in 1988 as maize fields (agriculture). On returning from the transect walk and reviewing the imagery, the author determined that there was no possible way, based on the satellite imagery, that the plot was previously under agricultural use. Having not obtained an aerial photograph from that year, the author pursued the issue with the community. During an informal community meeting near the spot, many individuals were insistent that the land "had always been used for maize fields, even in 1988." Even after having shown and explained the imagery and reconfirmed that the plot in the image was the contested plot on the ground, the local people were convinced that the plot was used for maize fields in 1988. The seeming conundrum was solved long after the meeting. The next week, while doing follow-up work in an area nearby, three people from the community meeting approached the author. They explained, in confidence, that the area had actually been grassland from the early 1980s until about 1996 or 1997. Several members of the community were occupying the land in hopes that the local chief and/or transitional local council would grant them permanent rights to the land. Part of the basis of their claim was that they had been consistently working the land for "many years." The informants maintained that if it was shown that the land was not used as recently as 1988, many feared that their claim to the land would not be accepted—thus the dispute over land use.

In this example, it is clear that a political ecology of land use, without the use of GIS, may have never identified the power struggle over this piece of land. Without making value judgments on the local representation of land cover (who is right, who is wrong), geospatial technologies helped to uncover "hidden political ecologies" in the landscape.

Summary of Case Study

Several themes emerged from the synthesis of GIS and political ecology in the study of land reform and changes in land cover/land use. First, a basis for the political ecology of land reform was formed. Both physical and social patterns were replicated across the CPAs. The political ecology of the land reform projects showed a clear reproduction of the historical relations of production. Labor was being extracted at high rates, leading to a lack of productive activity on the farms. The local and regional economies are still geared to providing migrant labor to the industrial centers, mines, and local commercial farms. Furthermore, credit continues to be constrained, especially for women. Outreach and extension are limited, and land for housing remains in greater demand than land for production. These factors overwhelm differences in environment at the local and provincial scale and are reinforced by structural problems found within the CPAs.

Second, through the investigation of the CPAs using a GIS, a broader regional political ecology was uncovered that directly impacted the political ecology of land reform. The key regional phenomenon that impacted the CPA included:

• *Poorly defined tenure relations.* Tenure reform has occurred, albeit in small steps. The tenure system is confusing and does not provide for widespread security. Women still experience a high degree of insecurity, especially in tenure. Authority over tenure needs to be clarified and codified in the former homelands, including the issuing of titles (Cross, 1988).

• *Overlapping and multiple claims to local authority.* South Africa's system of competing authorities leads to inconsistency and confusion. For instance, the membership of CPAs have to function in two roles, one on the CPA where authority is clear, and the other in their places of residence where authority is not always clear. The CPAs represent an alternative source of authority and as such threaten local chiefs. In fact, several chiefs were afraid that the CPAs would draw away participants' labor and tribute.

• *Gender inequities.* Redressing gender bias has been one of the central initiatives in the quest to transform South African society. Employment reform, access to health care, and equal status under the law have all been affected since 1994. However, women in the province have yet to feel the full force of these reforms.

• *Limited and unequal access to capital.* The availability of capital in the provincial economy is still restricted, with banks wary of investing in black agriculture. Credit is extended at shops for food and furniture, but rarely at banks for agricultural needs.

• *A skewed labor market and the continuing urban bias.* South Africa's labor market, geared toward providing cheap labor to the major industrial centers (Lipton, Ellis, & Lipton, 1996), dramatically effects the amount of labor in this rural province. Urban bias persists in South Africa (Lipton et al., 1996), for cities receive the larger portion of investment and consume the bulk of tax revenues.

• *A marginal and perilous environment.* Limpopo Province is not uniformly endowed with consistent and adequate rainfall, exceedingly good soils, or other environmental conditions necessary for high agricultural productiv-

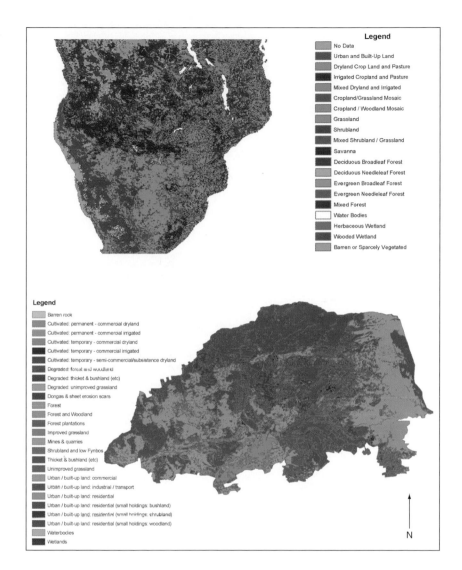

PLATE 10.1. Two representations of southern African land use and cover.

(top) Southern Africa land use and cover from NASA's Earth Observing System—Africa Land Cover Characteristic Dataset.

(bottom) Limpopo Province land use and cover from the Centre for Scientific and Industrial Research, South Africa.

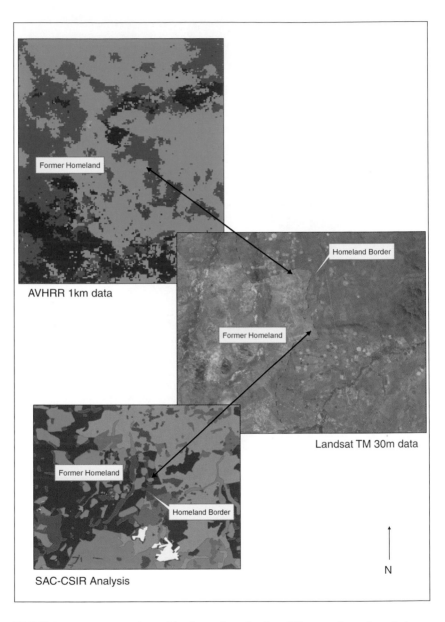

PLATE 10.2. Representations of the former homelands at different scales and resolutions.

PLATE 10.3. Land use and cover change on Mahlambandlovu CPA and environs: 1988–2000.

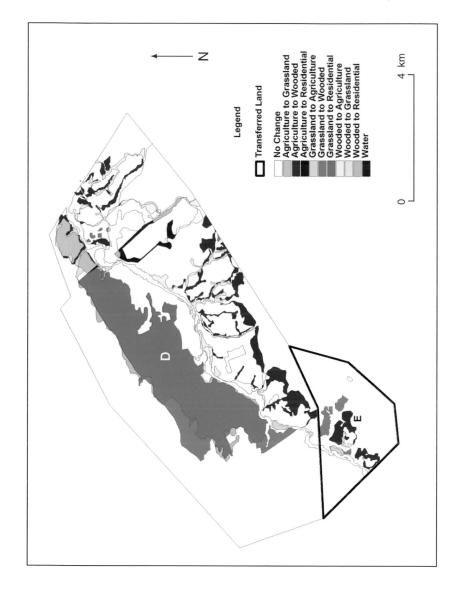

PLATE 10.4. Land use and cover change on Rondebosch CPA and environs: 1988–2000.

ity. Further, it is prone to floods and droughts that disrupt local livelihood systems and hinder economic development.

Third, socially and environmentally inappropriate land use patterns established under apartheid are being reproduced as "neo-apartheid" landscapes (Pickles & Weiner, 1991). For example, intensity of land use for agricultural production is not markedly higher in the higher productivity areas of the escarpment than in the significantly less productive highveld region. Only when land use patterns were examined off the transfer land did local political ecologies differentiate between environmental settings.

Fourth, the coarseness of data at the local scale made it difficult to understand the impact of the local environment at a very fine scale. Only through detailed construction of the household political ecology did environmental issues emerge at the plot and field level. The construction of this household and community information led to important lessons for comparison and for generalizing processes and patterns to the regional scale. In effect, the link between local and regional was forged in this exercise by integrating geospatial technologies with detailed and placed-based household and community information.

Finally, community and household narratives highlighted nature as an agent of its own, not only as socially produced. For example, in the area adjacent to the Mahlambandlovu CPA, agricultural intensification is taking place where there is access to irrigation water. In the Rondeboch CPA, the floods of 2000 in northern South Africa were a major problem to household stability and the ability to practice agriculture and had a significant impact on the land use and land cover change patterns.

CONCLUSION

We have argued in this chapter that global land cover and land use studies rely on geospatial technologies and produce representations of global/regional change that are poorly linked to human–environmental landscape dynamics. This was demonstrated with a Southern African case study where fundamental sociohistorical processes were subsumed within a regional-scale analysis of land cover and land use. As a result, the human–environmental dimensions of landscape change are assumed to be exogenous to the models produced and the representations of nature are overly data-driven. In the Southern Africa case, a simple GIS overlay of the former "homelands" in South Africa reveals how fundamental human dimensions of land cover and land use are not visible.

The Limpopo Province case study demonstrated how the techniques employed in producing a land cover/land use map and its classification can impact the spatial parameters of specific categories, and therefore representations of provincial scale on nature. At the local scale, geospatial technologies were integrated with community local knowledge to explore GIS–political ecology integration. Local knowledge about land use and land cover change helped to produce a more robust political ecology, while GIS helped to uncover "hidden political ecologies." This should not be a surprise, as GIS representations of land use and land cover are placed-based spatial stories for the participants who live there. For project participants, these classifications signified struggles for social reproduction and their attempts at post-apartheid reconstruction.

In the case study of land reform, the linking of GIS and political ecology was constrained by the spatial resolution of satellite images, limited availability of aerial photography for the study areas, temporally limited or inappropriate images (i.e., comparing rainy season to dry season), and limited temporal data on local ecology beyond basic maps. However, the integration worked well in allowing for assessment of land use and land cover change as described by project participants, the enhancement of participants' ability to visualize their actions in time and space, in augmenting GIS representations of nature, and in contributing technical data for a local political ecology analysis.

We expected to find a stronger relationship between environment and land use/land cover change in Limpopo Province. One explanation is the data resolution, which was too coarse. But common social forces are also at play in Limpopo Province, which seemingly overwhelm local environmental variations. In particular, apartheid and colonialism created a semiproletariat in the South African countryside that was spatially balkanized within the former homelands. With liberation in 1994, peoples' spatial strategies were significantly altered and land intensification has taken the form of rapid periurbanization in all ecological zones. In this way, the local informed the provincial and a GIS-based political ecology was useful for scaling up. Clearly, further investigation of the regional political ecology and the techniques with which we integrated local political ecology and GIS is needed.

REFERENCES

Aitken, S., & Michel, S. (1995). Who contrives the "real" in GIS?: Geographic information, planning, and critical theory. *Cartography and Geographic Information Systems, 22*(1), 17–29.

Blaikie, P. (1994). *Political ecology in the 1990s: An evolving view of nature and society* (CASID Distinguished Speaker Series No. 13). East Lansing: Michigan State University, Center for the Advanced Study of International Development.

Campbell, D. (1998). Toward an analytic framework for land use change. In L. Bergstrom & H. Kirschmann (Eds.), *Carbon and nutrient dynamics in tropical agroecosystems* (pp. 281–301). Wallingford, UK: CAB International.

Craig, W., Harris, T. M., & Weiner, D. (Eds.). (2002). *Community participation and geographical information systems*. London: Taylor & Francis.

Cross, C. (1988). Freehold in the homelands: What are the real constraints? In C. Cross & R. Haines (Eds.), *Towards freehold?: Options for land and development in South Africa's black rural areas*. Cape Town, South Africa: Juta.

Curry, M. R. (1995). Rethinking rights and responsibilities in geographic information systems: Beyond the power of image. *Cartography and Geographic Information Systems, 22*(1), 58–69.

Department of Land Affairs. (1997). *White paper on South African land policy*. Cape Town, South Africa: CTP Book Printers.

Gallopin, G. (1991). Human dimensions of global change: Linking the global and the local processes. *International Social Science Journal, 130,* 706–718.

Goss, J. (1995). "We know who you are and we know where you live": The instrumental rationality of geodemographic information systems. *Economic Geography, 71,* 171–198.

Harris, T., & Weiner, D. (1998). Empowerment, marginalization, and community-integrated GIS. *Cartography and Geographic Information Systems, 25*(2), 67–76.

Harris, T. M., Weiner, D., Warner, T., & Levin, R. (1995). Pursuing social goals through participatory GIS: Redressing South Africa's historical political ecology. In J. Pickles (Ed.), *Ground truth: The social implications of geographic information systems* (pp. 196–222). New York: Guilford Press.

Kwaku-Kyem, P. (1999). Examining the discourse about the transfer of GIS technology to traditionally non-western societies. *Social Science Computer Review, 17*(1), 69–73.

Lake, R. (1993). Planning and applied geography: Positivism, ethics, and geographic information systems. *Progress in Human Geography, 17*(3), 404–413.

Lambin, E. (1992). *Spatial scales, desertification, and environmental perception in the Bourgouriba region (Burkina Faso)* (Working Paper No. 167). Boston: Boston University, African Studies Center.

Levin, R., & Weiner, D. (1997). *No more tears: Struggles for land in Mpumalanga, South Africa*. Trenton, NJ: Africa World Press.

Lipton, M., Ellis, F., & Lipton, M. (1996). *Land, labour, and livelihoods in rural South Africa: Vol. 2. KwaZulu-Natal and Northern Province*. Durban, South Africa: Indicator Press.

Liverman, D., Moran, E. F., Rindfuss, R. R., & Stern, P. C. (Eds.). (1998). *People and pixels: Linking remote sensing and social science*. Washington, DC: National Academy Press.

McCusker, B. (2001). *Livelihoods and land use change in South Africa: The unfinished transformation*. Unpublished doctoral dissertation, Michigan State University, East Lansing.

Meyer, W., & Turner, B., (1994). *Changes in land use and land cover: A global perspective*. Cambridge, UK: Cambridge University Press.

Obermeyer, N. (1998). The evolution of public participation/GIS. *Cartography and Geographic Information Systems, 25*(2), 65–66.

Oppenshaw, S. (1991). A view on the GIS crisis in geography; or, Using GIS to put Humpty-Dumpty back together again. *Environment and Planning A, 23,* 621–628.

Pickles, J. (1991). Geography, GIS, and the surveillant society. *Papers and Proceedings of Applied Geography Conferences, 14*, 80–91.

Pickles, J. (Ed.). (1995). *Ground truth: The social implications of geographic information systems.* New York: Guilford Press.

Pickles, J., & Weiner, D. (1991). Rural and regional restructuring of apartheid: Ideology, development policy, and the competition for space. *Antipode, 23*(1), 2–32.

Rundstrum, R. (1995). GIS, indigenous peoples, and epistemological diversity. *Cartography and Geographic Information Systems, 22*(1), 45–57.

Sheppard, E., Couclelis, H., Graham, S., Harrington, J., & Onsrud, H. (1999). Geographies of the information society. *International Journal of Geographical Information Science, 13*(8), 797–823.

Sheppard, E., & Poiker, T. (Eds.). (1995). GIS and Society [Special issue]. *Cartography and Geographic Information Systems, 22*(1).

Taylor, P. J., & Overton, M. (1991). Further thoughts on geography and GIS. *Environment and Planning A, 23*, 1087–1094.

Turner, M. (2001). Merging local and regional analyses of land-use change: The case of livestock in the Sahel. *Annals of the Association of American Geographers, 89*(2), 191–219.

Weiner, D., & Harris, T. (2003). Community-integrated GIS from land reform in South Africa. Available from URISA Journal Online website: *http://www.urisa.org/Journal/onlinejournal.htm*

PART V

NORTH–SOUTH ENVIRONMENTAL HISTORIES

CHAPTER 11

Material–Conceptual Landscape Transformation and the Emergence of the Pristine Myth in Early Colonial Mexico

Andrew Sluyter

Bruno Latour's (1993) model of how the West (i.e., the North) has risen to global dominance over the Rests (i.e., the non-Wests, the Souths) in relation to the emergence of two conceptual dichotomies axiomatic to modernism provides the basis for understanding the allure and failure of orthodox international development (Sluyter, 2002, pp. 215–227). The first dichotomy separates nature from society (Figure 11.1). The second dichotomy externalizes the first to separate the Rests from the West. On one side, the Rests are a part of nature and therefore live in unchanging, undeveloped landscapes. They are premodern, habitually get nature and society mixed up, and thus remain locked in the grip of myths that naturalize social processes and socialize natural processes. On the other side of the West–Rests dichotomy, Westerners are apart from nature and therefore develop their landscapes while socially progressing. They are modern, carefully keep nature and society separate, and thereby achieve the objective understandings of each side of the nature–society dichotomy that will completely free society from nature's limits and society's members from each other—the so-called "double emancipation"

Adapted by Andrew Sluyter from the article "The Making of the Myth in Postcolonial Development: Material–Conceptual Landscape Transformation in Sixteenth-Century Veracruz," published in *Annals of the Association of American Geographers, 89*(3), 377–401 (1999). Adapted by permission of Blackwell Publishing. The adapting author is solely responsible for all changes in substance, context, and emphasis.

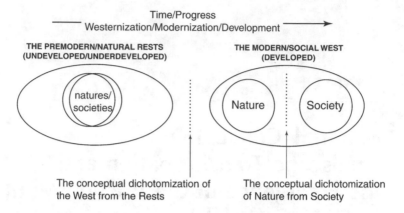

FIGURE 11.1. This diagram provides a basis for understanding the allure and failure of orthodox international development by modeling how the West (i.e., the North) came to perceive the Rests (i.e., the non-Wests, the Souths) as it rose to global dominance over the last several centuries. One major conceptual dichotomy that has emerged separates nature from society; the other separates the Rests from the West. Orthodox development theory assumes that the heterogeneous Rests must undergo a process of economic development, social modernization, and cultural Westernization to emulate the homogenous West's belief in the nature–society dichotomy and thereby cross from one side to the other of the West–Rests dichotomy. This figure is based on Latour (1993, p. 99).

(Sluyter, 2003). Orthodox development theory assumes that the heterogeneous Rests must go through the processes of economic development, social modernization, and cultural Westernization to cross from one side to the other of the West–Rests dichotomy, emulate the homogeneous West's belief in the nature–society dichotomy, and work toward the double emancipation.

As is all too clear more than half a century after the Bretton Woods Conference, orthodox development has been creating the very problems it claims to be solving (Escobar, 1995; Sachs, 1992). Beguiled by the promise of the double emancipation, the Rests have increasingly crossed the West–Rests dichotomy and joined Westerners in acting as if environmental limits would fade away with progress toward the telos of the double emancipation (Rosenberg, 1982). But by now only the incredibly ignorant or the incorrigibly self-interested would deny that development's boomerang effects have proliferated to a counterproductive degree, particularly natural–social phenomena of global consequence such as the greenhouse effect and green revolution crops reliant on disappearing fossil water and fuel. Such global natural–social hazards now threaten society more than any natural hazard ever did while stymieing improvements in social well-being (World Watch Institute, 2002).

The realization that orthodox development has resulted in a tragic declension toward environmental disaster and social disparity has eroded but not eliminated the nature–society and the West–Rests conceptual dichotomies. One consequence has been that the so-called traditional ecological knowledges (TEK) of the Rests have gained some acceptance as sustainable, productive alternatives to orthodox development—and not just among iconoclasts like Carl Sauer (1956), but among development institutions (United Nations, 1993, p. 227). Most basically, the Rests' emphasis on dynamic, heterogeneous landscapes would usefully complement the West's emphasis on stable, homogeneous landscapes (Zimmerer, 1996). Yet the diffusion of institutions and technologies from the West to the Rests and the counterflow of capital continues to dominate international development (Peet & Watts, 1996). The heterogonous knowledges of the Rests continue to be devalued as backward and static, as reflected in the word "traditional" that leads TEK. Much development based on TEK thus continues to implement homogenous Western objectives by coopting and decontextualizing selected aspects of knowledges specific to unique places, eliminate their dynamism, and focus more than anything else on negotiating the terms for their commodification (Escobar, 1998; Sluyter, 2002, pp. 204–209).

Nonetheless, the erosion of the two conceptual dichotomies has emphasized the need for research on the processes through which they arose in concert with the West qua West, including research on such crucial aspects of that process as "the pristine myth." That misconception is a necessary aspect of the West–Rests dichotomy, and therefore of orthodox development, because the pristine myth categorizes the precolonial landscapes of the Americas as undeveloped, as natural, as "wilderness" in the sense of uninhabited land (Denevan, 1992). Known globally as "the myth of emptiness" (Blaut, 1993, p. 15), the pristine myth is axiomatic to the pro-modern desire to develop supposedly unexploited resources as well as to the anti-modern desire to preserve supposedly unspoiled wilderness. Several generations of historical ecologists have conducted sufficient research to demonstrate that the precolonial Americas were in many places densely settled and intensively used, yet the pristine myth continues to support the West–Rests dichotomy and thereby to counterproductively influence development policy (Turner & Butzer, 1992). Since the pristine myth is a foundational element in the conceptualization of the West, only a much better understanding of how that misconception emerged and has persisted despite so much contrary evidence can debunk it and thereby facilitate moving beyond orthodox development.

Understanding the emergence of the pristine myth requires acknowledging and addressing two of its fundamental characteristics.

First, its emergence must have entailed simultaneous material and conceptual landscape transformations. The work of urban geographers demonstrates that landscape acts as a "visual vehicle of subtle and gradual inculcation ... to make what is patently cultural appear as if it were natural" (Duncan, 1990, p. 19). In other words, landscape patterning is both a result of transformation and a parameter for further transformation through processes such as human labor and categorization, with landscape patterning influencing both the habits of thought and practice that lend regularity to such processes, as well as impacting the conflicts that disturb that regularity. As conceptual transformations turned *productive* precolonial landscapes into *pristine* precolonial landscapes, material transformations must have simultaneously been convincing people that those landscapes really were pristine. Better understanding of the pristine myth will therefore require analytical integration of material and conceptual transformations (i.e., conceptual–ecological integration).

Second, material–conceptual landscape transformation must involve processes that span the nature–society dichotomy. Cronon (1983) has suggested that recategorization of the New England landscape as pristine was a result of native depopulation due to epidemic diseases and consequent contraction of native settlement, reduction of regular burning of forests, and vegetation succession to a denser, darker forest that appeared to be pristine. Yet the nature–society dichotomy dictates that scientific disciplines study either natural *or* social phenomena (Sluyter, 2003). Consequently, responsibility for the proliferation of natural–social phenomena such as transgenic organisms falls to many disciplines while responsibility for understanding their boomerang effects falls to none. Political ecology has itself been oriented mainly toward social processes, from a foundation in political economy to the recent integration of culture as an integral rather than a supporting aspect of social change (Bryant & Bailey, 1997; Peet & Watts, 1996; Stott & Sullivan, 2000). The political ecology research most directly related to the emergence of the pristine myth has emphasized the colonial reconfiguration of the global political economy and treated the concurrent emergence of categorizations such as pristine wilderness as ideological props for Western dominance (Blaut, 1993; Wolf, 1982). Yet because such global categories refer to local landscapes that changed in part through biological aspects of the colonization process (Crosby, 1986), better understanding of the pristine myth requires that political ecologists incorporate analysis of phenomena such as vegetation change (i.e., political–ecological integration) and link analyses of local landscapes to analyses of global political economy, environment, and discourse (i.e., scale integration).

In order to gain insights into the emergence of the pristine myth, this chapter integrates analysis of some social and biological processes involved in a material–conceptual landscape transformation early in the colonial period. The case study focuses on the Veracruz lowlands during the first century of colonization, from 1519 to 1619, because that landscape served as the beachhead for Spanish colonization of the mainland and thus influenced the initial Western conceptualization of the landscapes of New Spain as well as undergoing some of the earliest material transformations due to disease and livestock introductions (Figure 11.2). A systematic database of land grant documents enables partial reconstruction of interactions among vegetation, settlement, agriculture, ranching, and categories of land use, land cover, and land tenure.[1]

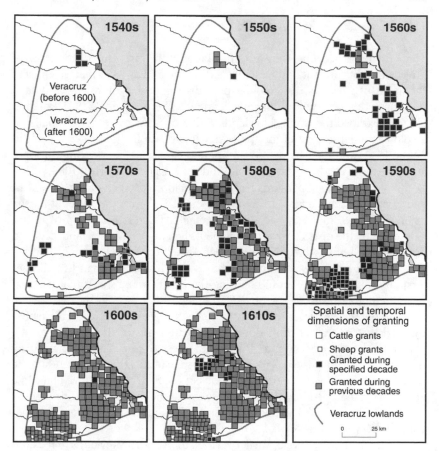

FIGURE 11.2. A database of land-grant documents enables reconstruction of the spread of colonial cattle and sheep *estancias* through the Veracruz lowlands between 1540 and 1619. While Spaniards introduced the first cattle to New Spain in the 1520s, the first land rush engulfed the narrow coastal plain only in the 1560s, prompted by a silver mining boom and associated population increase. Another land rush, focusing on the southern piedmont while filling in gaps on the coastal plain, occurred in the 1580s and 1590s. The effective cessation of granting by 1619 paralleled the decline in silver production and modest population growth during the colony's second century.

THE VERACRUZ LOWLANDS

Three firsthand accounts—by a conquistador at the inception of the colonial project, by a bureaucrat of the established colony, and by a German scientist at the dawn of Mexico's postcolonial period—signpost a dramatic landscape transformation. A few months after landing at Veracruz in 1519, Hernando Cortés described the tropical lowlands with a possessive eye, conceptualizing a colonization prospectus for the sovereign to whom he wrote: "beautiful bottomlands and river banks," "very apt and agreeable for traveling through" and, prospectively, "for pasturing all kinds of livestock" (Cortés, 1988, p. 20). Within a century the Spaniards had indeed colonized much of the Americas, and Cortés's beachhead had become the port of Veracruz, the entrepôt for the colony of New Spain. They had also implemented Cortés's pastoral prospectus, as attested by the district governor in 1580 when he described a landscape "so fertile and full of pastures that more than 150,000 head of livestock, between the cows and the mares, ordinarily graze within little more than [29 kilometers] all around, even without counting the innumerable sheep that descend from the highlands to overwinter" (JGI, ms. xxv-8, f. 5).[2] Yet by 1850, in the first comprehensive description of the region's vegetation, Carl Sartorius bemoaned that once past the woodlands and wetlands of the narrow coastal plain much of the region was a "dreary wilderness, overgrown with low thorny mimosas" (Sartorius, 1961, p. 9). The only respite for the traveler ascending the piedmont toward the temperate elevations of the Sierra Madre Oriental came when the road dipped into the verdant gorges of the major streams. Somehow, between 1580 and 1850, a landscape "so fertile and full of pastures" had become "dreary wilderness."

When orthodox development overtook the Veracruz lowlands in the 20th century, that pristine myth still prevailed. Until the federal government began to build irrigation projects in the 1930s, the region had remained little cultivated and largely unsettled, the domain of vast cattle herds (Sluyter, 2002, pp. 191–201). According to the teleological model of the planners, irrigation agriculture would supplant ranching just as inevitably as ranching had supplanted the supposedly pristine precolonial landscape, and the Veracruz lowlands would make their definitive crossing from one side to the other of the West–Rests dichotomy. Except that if the planners had not been so myopically focused on homogenizing hydrology when studying their large-scale aerial photographs, they might have noticed the diagnostic patterns of precolonial settlement and intensive agriculture (Sluyter, 2002, pp. 38–60). They might have realized that the precolonial landscape had been every bit as densely settled and intensively cultivated as anything they planned for postcolonial Veracruz.

Most basically, then, a local version of the pristine myth had already emerged by the early postcolonial period, when Sartorius conceived of the landscape as a wilderness. He recognized some of the vestiges of precolonial settlement and agriculture but nonetheless argued for Westernization on the basis of the introduction of exotic crops such as coffee and sugar cane (Sluyter, 2002, pp. 189–191). The 20th-century planners merely succeeded in implementing a more grandiose version of the project that he never managed to bring to even modest fruition.

But if the issue of *when* the pristine myth emerged seems relatively straightforward, *how* it emerged remains a more complex issue. Considering the extremes of the range of possibilities provides a useful way to begin sorting out what happened.

At one extreme, several centuries of overgrazing and degradation of what was prime range in 1519 might have resulted in a "dreary wilderness" by 1850. That lowland Veracruz actually had as much lush grassland in 1519 as implied by Cortés's description is fairly certain despite its brevity, boosterism, and locational ambiguity—perhaps extrapolating to the entire region from the verdant environs of Zempoala, a city of some 100,000 on the narrow coastal plain. A 1529 letter sent home by a less famous colonizer confirms that lush grasslands extended well beyond the coastal plain: "There are [42–84 kilometers] of plain as flat as a floorboard, and in some places [168–336 kilometers], with grass as high as your knee or higher" (quoted in Lockhart & Otte, 1976, pp. 195–196). The 1580 report also confirms an extent of grasslands sufficient to support 150,000 head of large stock and "innumerable" transhumant sheep (JGI, ms. xxv-8, f. 5). That by 1850 those grasslands had become overgrown with thickets is also fairly certain because savanna with extensive patches of deciduous shrubland and woodland just like that which Sartorius described still dominates except where obscured by irrigation agriculture.

Yet such descriptions are too brief, given to exaggeration, and unsystematic to rule out other possibilities, even the opposite extreme of the range of possibilities, namely, that the seemingly degraded grasslands Sartorius noted in 1850 might already have existed when the Spaniards arrived—the result of precolonial land use, climate, or both. After all, Cortés and many of his compatriots hailed from Estremadura and thus took one of the dustiest corners of Spain as their measure of what was prime rangeland and what was not. A landscape they perceived as verdant might well look like "dreary wilderness" to a northern European like Sartorius. Moreover, extreme overgrazing during the colonial period would have resulted in gullying and sheet erosion, for which no evidence exists (Sluyter, 2002, pp. 118–119).

Regarding precolonial land use, pieced-together estimates of native population reveal that a dense agricultural population occupied these lowlands in 1519, with some 100,000 Totonac at Zempoala and a regional population in excess of half a million (Sluyter, 2002, pp. 38–47). All those people produced their own food plus a surplus—mainly cotton tribute for the Aztec Empire. Relict earthen mounds, stone pyramids, sloping-field terraces, and the ditches of intensive wetland agriculture still litter the landscape and attest to the density of that occupance—at least to those who actively search out such faint spoor of ancient labor (Sluyter, 2002, pp. 38–60). Although the chronologies of many of those vestiges remain uncertain, the sediments of a small lake on the coastal plain preserve a pollen record of maize agriculture several millennia long (Sluyter, 2002, pp. 58–59). Precolonial agriculturalists, long and densely settled on the landscape, would have had ample opportunity to impact the vegetation.

Regarding climate, the vegetation that Sartorius describes might derive, at least in part, from the long winter dry season. The upper piedmont and the escarpment rising inland to the peaks of the Sierra Madre receive the greatest rainfall, with the trade winds condensing as they sweep upslope (García, 1970). But a rainshadow parches the region from November through April, particularly in the north and on the lower piedmont. How climate might have changed in these lowlands over the last 500 years remains unknown, but some data indicate that the Little Ice Age might have expressed itself in Mexico as a drier-than-present period that began in the second half of the 16th century and did not abate until about 1850 (Florescano, 1980). Regardless, even the present climate could account for low deciduous forest and, where vertisols inhibit tree growth, a patchwork of savanna and shrubland (Gómez-Pompa, 1973).

Between the extremes of the range of possibilities lies a complexity about which we know approximately nothing, even focusing on vegetation change let alone considering conceptual transformation as well. Fortunately, the systematic database of land grants enables a partial reconstruction of pastoral land use and vegetation, subsequent analyzes of material and conceptual landscape transformations, and their integration into an analysis of material–conceptual landscape transformation.

RECONSTRUCTION OF PASTORAL LAND USE AND VEGETATION

While Gregorio de Villalobos introduced the first cattle soon after Cortés wrote his prospectus for colonization, the viceroy began to grant *estancias* only after two decades of informal, small-scale ranching (Sluyter, 2002, pp. 67–79). And, even with that opportunity to secure land tenure, the first land rush did not occur until the 1560s, when the mining boom and consequent increase in Spanish immigration prompted a flood of grants that stretched the length of the narrow coastal plain—although some of those grants might have formalized preexisting *estancias*. Another land rush in the 1580s—as the native population of New Spain approached its nadir after the epidemics of the previous decade, resulting in reduced food supplies, high prices, and land grabbing—focused on the southern piedmont while filling in gaps on the coastal plain.

The continuing growth of the textile industry and a consequent need for low-land winter pasture prompted the flood of sheep grants in the 1590s. The effective cessation of granting by 1619 paralleled the decline in silver production and modest population growth of the colony's second century. Besides, except for the relatively inaccessible central piedmont, the landscape was full. Some 300,000 head of cattle and sheep, and possibly an astounding five times that many, grazed more than half of the lowlands inland from the port of Veracruz (Sluyter, 2002, pp. 103–113).

Beyond reconstructing the spatial and temporal dimensions of the livestock invasion, the same documents permit a systematic evaluation of its relationship to vegetation change. In an attempt to ensure that grants would not prejudice the interests of the Crown, other ranchers, or native communities, a viceregal official would inspect the location of a prospective *estancia* and submit a report that included a recommendation to award or not award the grant—although almost invariably the former. Sometimes those reports include descriptions of terrain, vegetation, and settlement. Mapping the vegetation references yields a perspective on the vegetation of a particular place during the livestock invasion and thereby allows an analysis of a major element of material landscape transformation (Figure 11.3).

MATERIAL LANDSCAPE TRANSFORMATION

As granting spread southward during the 1560s, the vegetation references associated with the coastal plain refer to savannas with *matas, montes*, and *matas de monte*: a matrix of grassland with patches of fragmented, open woodland or shrubland. As one inspector put it in 1574, there was "nothing except *montes* and savanna" (AGN-T, vol. 32, exp. 4). The subsequent expansion into the dunes reveals a similarly open vegetation of grass and thickets. On the piedmont, the inspectors encountered stands of evergreen oaks at higher elevations but also the familiar savannas with areas of deciduous open woodland and shrubland. Throughout the region, as at present, evergreen trees fringed rivers and seasonal wetlands, and mangroves bordered the brackish coastal lagoons.

At the regional scale those patterns echo Sartorius's description and the current vegetation in those places not obscured by irrigation projects; therefore, the colonial livestock invasion must have occupied savanna with deciduous thickets rather than degrading more lush, continuous grasslands. In its broadest pattern, that vegetation seemingly follows climatic and edaphic variation. Shrubs and herbs dominate much of the coastal belt of dunes in part because the sand substrate, despite receiving more than 1,000 millimeters of precipitation annually, precludes even low deciduous forest except on the finer substrates of interdunal basins (García, 1970). A patchwork of savanna and low deciduous woodland occurs on the upper piedmont, despite more than

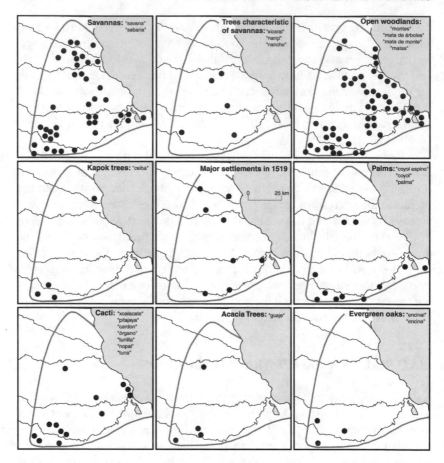

FIGURE 11.3. The database of land-grant documents also enables reconstruction of the vegetation that the livestock invaded. Vegetation types such as savannas and particular species such as kapok trees reveal colonization's impact on the landscape. The livestock invasion at most modulated the balance between the herbaceous and the woody species during an overall process of vegetation change precipitated by sudden and extreme depopulation of what had been a densely settled and productively used precolonial landscape. For reference, the central panel maps the major settlements of the Totonac and Nahua peoples who dominated the region when the Spaniards arrived in 1519.

1,200 millimeters of precipitation, in part because the dominant vertisols preclude tree growth due to waterlogging during the wet season and deep cracking during the dry season. The low deciduous woodland grades into shrubland on the drier lower piedmont. Only the wetlands of the coastal plain harbor extensive evergreen communities. On the northern coastal plain, vertisols and the rainshadow suggest savanna with sparse tree growth as the precursor to the irrigated cane fields that have replaced most other vegetation. On the southern coastal plain, increasing precipitation and phaeozems also suggest savanna but with the possibility of extensive low deciduous woodland and shrubland.

Scrutinizing more specific vegetation references, the grants also reveal that precolonial land use had locally affected the vegetation that the livestock invaded. As native population rapidly declined in the first two decades of colonization, from some 500,000 to less than 100,000 after the initial smallpox and measles epidemics, characteristic plant species invaded former settlements and agricultural fields.

The grant inspectors frequently noted two species of palm tree, the coyol espino (*Acrocomia mexicana*) and the coyol (*Scheelea liebmannii*), both of which characteristically invade disturbed vegetation. Because the coyol bears edible fruit, the natives even encouraged its growth, the 1580 geographical report classifying it as a "cultivated tree of this land" (JGI, xxv-8, f. 11r). In several cases, the grants clearly associate coyol palms with savannas near former native settlements (AGN-T, vol. 3331, exp. 1, ff. 1r–9r; AGN-M, vol. 33, ff. 112v–114r, 115v–116v). One grant even notes "a savanna where there is a round *mata* that in the middle of it has a large clearing with two palms" (AGN-M, vol. 14, ff. 80v–81v)—a striking, but at the time seemingly unappreciated, description of thickets invading former agricultural fields.

If the coyol palm marked moribund agricultural fields, the *ceiba* marked the former settlements themselves. Natives protected the ceiba, the kapok or silk-cotton tree (*Ceiba pentandra*), because its enormous canopy provided shade, its seed pods fiber, and its bark medicine. Sacred among the Maya as the "World Tree," towering *ceibas* still grace the plazas of towns throughout the lowlands. The grants note *ceibas* near Zempoala as well as more generally associated with the stone and earthen mounds, the so-called *cúes*, that marked former settlements (AGN-M, vol. 9, ff. 33r–33v; vol. 10, ff. 45v–46r; vol. 14, ff. 161v–162v; vol. 15, ff. 25v–26v, 86r–87r; vol. 17, ff. 25v–26v). Significantly, lone *ceibas* rose above thickets of low trees and shrubs: "in a savanna where there is a large mata, and in the middle of it a large *ceiba* rising above the trees of the said mata"; "a round mata and in it a very large *ceiba*" (AGN-M, vol. 15, f. 26r; vol. 17, ff. 25v–26r). The *ceibas*—and, more significantly, the *matas de monte* in which *ceibas* occurred—marked the overgrown settlements of the former native population, signposts to a moribund cultural landscape.

By the time of Sartorius, three centuries later, the "dreary wilderness, overgrown with low thorny mimosas" that he describes was the outcome of a struggle between grasslands and thickets that would have hinged on the balance between disturbance by livestock versus disturbance by fire. Heavy grazing would have promoted woody thickets because it reduces the supply of fine fuels necessary to frequent low-intensity fires that suppress the seedlings of woody plants. At the same time, browsing by cattle of the leguminous pods of mimosas disperses their seeds in fertilizing cow flops and thus further promotes thicket invasion. In contrast, repeated active burning of pastures by ranchers would have consumed the woody species and favored the grasses. Despite viceregal prohibitions, such burning seems to have been regular practice during the 16th century, with one Veracruzan reporting that the pastures were "wont to be burned around Christmas time" and sedimentary charcoal from Laguna Catarina confirming the prevalence if not the seasonal timing of burning (Paso y Troncoso, 1905, vol. 5, pp. 194–195; Sluyter, 2002, pp. 118–121). Such December burns, relatively early in the dry season when vegetation is still relatively incombustible, would have resulted in localized low-intensity fires that fragmented the landscape into a mosaic of grassland and woodland patches. The practice of burning late in the dry season that Sartorius (1961, p. 81) implies predominated by the 19th century would have resulted in much more extensive and intense fires that created more homogenous grasslands. If the Little Ice Age did cause drier or shorter wet seasons than during the precolonial and postcolonial periods, the herbaceous would only have been further favored.

While much remains uncertain about that process and the state of the vegetation at any particular time and place, the land grants yield a more systematic view of material landscape transformation than descriptions by individual Spaniards can ever provide. Neither extreme in the range of possibilities occurred: a degraded landscape did not already exist in 1519, nor did colonial overgrazing degrade a lush precolonial landscape.

Instead, the descriptions by Cortés, the district governor, and Sartorius reflect a more ambiguous process. Cortés and his conquistadors rode through a landscape that manifested millennia of resource management through the burning of vegetation, hydraulic engineering, and erosion control: patches of settlement and agriculture in a matrix of cultural savanna, as opposed to a purely climatic or edaphic savanna. But by the time of the land rush of the 1560s, the patches of former settlement and agriculture with *ceibas* and palms marking the graves of thousands were undergoing old-field succession, the thickets having invaded before the livestock. Sartorius rode through a landscape that three subsequent centuries of grazing and range management practices such as burning and, possibly, climate change, had further transformed. But those processes had at most modulated the balance between the herbaceous and the woody during an overall process of vegetation change precipitated by sudden and extreme depopulation of what had been a densely settled and productively used precolonial landscape.

CONCEPTUAL LANDSCAPE TRANSFORMATION

The Spaniards not only diffused epidemic diseases that precipitated a material landscape transformation, they implanted categories that precipitated a conceptual landscape transformation. In particular, they deployed a particular model of the relationship between land use and land tenure in combination with a failure to understand, recognize, or even consider native categories of land use. That fatal combination of a normative vision with a failure to see what was recategorized agricultural lands beyond the immediate environs of surviving native communities as unused pasture.

The grant inspectors characteristically described prospective *estancias* as *baldíos* (AGN-M, vol. 8, ff. 190–190v; vol. 9, f. 91v; AGN-T, vol. 2702, exp. 12, ff. 386–397v; vol. 2702, exp. 13, ff. 398–406v; vol. 2764, exp. 15, ff. 181–195v; vol. 2777, exp. 3, ff. 1–9v; vol. 3331, exp. 1, ff. 1–9). The term *baldío* equates to "wasteland," similar to the *wasta est* that described idle lands in the 11th-century Domesday Book of England (Darby, 1973). Both *baldío* and wasteland designate lands once productive but made idle by plague or war. In England, the Norman Conquest had laid waste such lands. In Iberia, the Reconquista turned Muslim lands into *baldíos* that the Crown then granted to Christian settlers. The privilege to graze lands not actively used extended even to crop stubble after the harvest, an arrangement in which transhumant flocks gained access to valuable fodder and left behind an equally valuable deposit of manure. The term *baldío* thus did not categorize lands as pristine, as lacking any prior productive use, but as idle due to depopulation.

In so frequently applying the category *baldío*, then, the grant inspectors recognized that epidemic disease had laid waste the Veracruz lowlands. Yet while the inspectors often noted the *cúes* of moribund settlements, the associated vestiges of agricultural fields went entirely unremarked. The Spaniards seem to have been unaware that natives had once excavated a labyrinth of ditches to manage the hydrology of the wetlands of the coastal plain, although the vestiges of more than 2,000 hectares of such ditches and intervening planting platforms remain discernible on aerial photographs (Sluyter, 2002, pp. 56–59). Only one grant inspector noticed anything at all: "a small lake which appears in the rainy season . . . and marshes ditched straight southward" (AGN-M, vol. 15, ff. 191r–192r). Not a single inspector noted the even more extensive sloping-field terraces of the lower piedmont, also still visible from the air (Sluyter, 2002, pp. 51–56).

Lands that natives clearly and actively used did not merit categorization as *baldíos* or granting as *estancias*, but conflict arose when ranchers insisted on seasonal access to crop stubble. Elsewhere in New Spain, ranchers went so far as to sue natives for fencing their maize fields to exclude cattle (AGN-M, vol. 3, f. 328). Some of that conflict arose from a failure to see essential differences between Spanish and native agroecologies. Wheat scythed at harvest leaves only the butts of stalks in the fields. Livestock can graze that stubble without damaging the crop while leaving behind manure essential to the next crop. Maize undergoes a much more protracted harvest, however. At the end of the growing season, the stalks sometimes are doubled over without removing the ears. The inverted husks shield the ears while drying. Even fields of yellowed, doubled-over stalks, which a Spaniard might have equated with the stubble of a harvested wheat field, could retain a substantial number of drying ears. While ranchers won their suits and access to harvested maize fields, it so threatened native subsistence that by 1560 legislation had restricted access to January and February, well after the end of the maize harvesting period (Chevalier, 1952, p. 119).

The case of Espiche in the early 1570s illustrates another, more insidious process—one that pertained to landscape patches such as fallow fields and diverse resource-gathering zones that fit more ambiguously into the Spanish conceptualization of the relationship between land use and land tenure. The grant inspector in that case claimed that the natives of Espiche themselves categorized the prospective *estancias* as *baldíos* (AGN-T, vol. 32, exp. 4). But a lawsuit prompted an investigation that revealed that some of the lands being granted were in reality native orchards and agricultural fields. They might have been fallow at the time of inspection and thus overgrown with grass and thickets, but they nonetheless remained part of the native subsistence system. Moreover, the rancher requesting the *estancias* had used alcohol to coerce the natives into agreeing not to contest the categorization as *baldíos*. Fallow fields, orchards, and communal resource-gathering zones also existed in Spain, of course, but when confronted with unfamiliar agroecologies and an exuberant tropical vegetation, the category of *baldío* became exceptionally plastic. Apparently its application to fallow lands also occurred in other regions of New Spain (Chevalier, 1952, p. 279; Rojas Rabiela, 1988, pp. 64–65).

Such categorization of cultivated and otherwise managed lands as *baldíos* not only underrepresented the extent of precolonial agricultural fields, many undergoing old-field succession, but thwarted viceregal legislation intended to protect the native communities that had survived the epidemics of the first half of the 16th century. Nonetheless, the viceroy allowed the Espiche grants to stand, and that community did not survive the Great Cocolixtle epidemic of the late 1570s. Nor, weakened in part by such depredations, would more than a handful of other native communities survive in the Veracruz lowlands (Sluyter, 2002, pp. 153–159). And even those settlements were so diminished that a Jesuit would claim, albeit somewhat hyperbolically, that by the 1620s not a single native was to be seen within 67 kilometers of the port of Veracruz (Pérez de Rivas, 1896, vol. 2, p. 195).

Thus, in accumulating space, at least some Spaniards circumvented the legislation intended to protect native land tenure and thereby helped to make such legislation superfluous—on the books but with few native communities left to protect by the end of the first century of colonization. More significantly, in circumventing laws, individuals were able to draw on taken-for-granted environmental and spatial categories that formed the conceptual parameters for colonial space accumulation. Lands beyond the immediate confines of native communities equaled wasteland, a patchwork of seemingly unused grasslands and shrublands most rationally suited for grazing. Space equaled a map, the maps associated with some of the land grants at once recording space accumulation and land use while manifesting the conceptual parameters for the articulation of power through space in order to exert power over space. As Harley (1992, p. 532) notes, "Cartography is part of the process by which territory becomes." While natives also had cartographic traditions, their modes of representation differed so much from those of the Spaniards that native maps did not provide an immediate basis for resistance (Sluyter, 2002, p. 183).

MATERIAL–CONCEPTUAL LANDSCAPE TRANSFORMATION

To summarize the material and conceptual landscape transformations while integrating them into a single material–conceptual analysis, the interaction of precolonial land use and climate created a matrix of cultural savanna with patches of settlement and agriculture. That material landscape patterning together with landscape concepts and laws diffused from Spain through the Antilles constituted the parameters for Cortés's vision of a pastoral landscape. Depopulation due to epidemics of introduced diseases created a moribund landscape, with thickets invading former fields and settlements before the livestock. As the ranchers accumulated space and increasingly occupied the landscape with their herds and flocks after midcentury, they preempted the recovery of the surviving native population. Epidemiological and old-field successional processes thus resulted in a material landscape pattern that together with the conceptual landscape pattern inherent in such categories as baldío constituted the parameters for further landscape transformation. The recategorization of moribund as well as fallow fields and orchards as baldíos obscured the native labor that had created those landscape patches, that would have provided the legal basis to prevent dispossession, and that would thus have made possible the recovery of the native population and reversal of old-field succession.

In effect, a positive feedback loop linked the material and conceptual transformations and rendered them inexorably unidirectional. Recategorization of old-field and fallow-field landscape patches as wasteland preempted the recovery of the native population, positively stimulating further old-field succession, and further recategorization of landscape as wasteland. The result of such positive feedback was a runaway transformation into a depopulated landscape of livestock *estancias*. The land use, land cover, and land tenure categories manifest in the land grant documents became self-ratifying, materially precipitating the very landscape they erroneously described by visually validating themselves and erasing precolonial categories. The result was a myth of progress from a supposedly pristine landscape toward a productive landscape when, in fact, the opposite was occurring. By the time the 20th-century development planners arrived on the scene, the material–conceptual landscape transformation had progressed to the degree that they were completely unaware of the productive agriculture of the precolonial period and had no clue that it might provide sustainable alternatives, rooted in and integrated with that particular place for centuries, to an irrigation project to grow exotic crops such as sugar cane.

CONCLUSIONS

Ultimately, the issue is not that a few Spaniards manipulated land use categories to circumvent legislation intended to protect native communities. It is not simply that native communities, devastated by epidemic diseases and other aspects of the colonization process, have taken time to learn how to use those land use categories to resist further dispossession. Ultimately, the issue is that those categories became the taken-for-granted, be-all and end-all measures of productive land use and continue to thwart effective development policy. Ultimately, the issue is to understand how those categories became so dominant and doggedly remain so through a material–conceptual feedback process that not only resulted in the pristine myth but that simultaneously obscured the emergence of the pristine myth qua myth. Such understanding is central to learning how to appreciate the landscapes of the Rests—from the many precolonial ones to the ever-diminishing (post)colonial ones—as alternatives to orthodox development.

That a similar feedback process might have operated in other regions of the Americas, such as New England (Cronon, 1983), suggests the possibility of modeling the emergence of the pristine myth for different types of colonization and thereby better understanding it (Sluyter, 2002, pp. 211–215). In terms of political ecology, settler and franchise colonization would each have entailed distinct interactions among natives, colonizers, and landscape—the so-called "colonial triangle" (Sluyter, 2002, pp. 10–11). Settler colonization is associated with a landscape transformation that removes the natives and accumulates space for the colonizers (Wolf, 1982). Franchise colonization is associated with a landscape transformation that exploits native or transported labor and involves little settlement by the colonizers. The pristine myth would therefore have emerged through different material–conceptual transformations in settler and franchise colonies. In terms of regional political ecology, long isolation from the major hearths of animal domestication distinguishes the Americas from Africa or Asia in terms of susceptibility to epidemic diseases (Crosby, 1972); and biotic and agroecological differences distinguish temperate from tropical colonization (Crosby, 1986). The period of colonization (Souza, 1986)—affecting technological, conceptual, and other parameters—suggests a further basis for classifying types of colonialism in order to model the emergence of the pristine myth. Each of the models would need to analytically integrate material and conceptual transformations, pertinent social and biophysical processes, and linkages between changes in landscapes and global political economy, environment, and discourse.

ACKNOWLEDGMENTS

Although the views expressed in this chapter do not necessarily reflect those of the people acknowledged, I thank Karl Zimmerer and Tom Bassett for inviting me to contribute this chapter and Frances Hayashida and William Doolittle for offering feedback on drafts. The cartography laboratory of the Department of Geography of the University of Wisconsin at Madison reworked Figures 11.2 and 11.3.

NOTES

1. A longer presentation of this case study—including full citations, discussion of methods, and explication of backward and forward theoretical linkages—can be found elsewhere (Sluyter, 2002).
2. The following archival abbreviations are used throughout: JGI, Joaquin García Icazbalceta Collection of the Benson Latin American Library, Austin, Texas; AGN-T, Tierras section of the Archivo General de la Nación, Mexico City; AGN-M, Mercedes section of the Archivo General de la Nación, Mexico City

REFERENCES

Blaut, J. M. (1993). *The colonizer's model of the world: Geographical diffusionism and Eurocentric history.* New York: Guilford Press.
Bryant, R., & Bailey, S. (1997). *Third world political ecology.* New York: Routledge.
Chevalier, F. (1952). *La formation des grands domaines au Mexique: Terre et société aux XVI–XVII siècles.* Paris: Institut d'Ethnologie.
Cortés, H. (1988). *Cartas de relación.* Mexico City: Editorial Porrúa.

Cronon, W. (1983). *Changes in the land: Indians, colonists and the ecology of New England*. New York: Hill & Wang.

Crosby, A. W. (1972). *The Columbian exchange: Biological and cultural consequences of 1492*. Westport, CT: Greenwood Press.

Crosby, A. W. (1986). *Ecological imperialism: The biological expansion of Europe, 900–1900*. Cambridge, UK: Cambridge University Press.

Darby, H. C. (1973). *Domesday Book: The first land utilization survey*. In A. R. H. Baker & J. B. Harley (Eds.), *Man made the land: Essays in English historical geography* (pp. 37–45). Newton Abbot, UK: David & Charles.

Denevan, W. M. (1992). The pristine myth: The landscape of the Americas in 1492. *Annals of the Association of American Geographers, 82*, 369–385.

Duncan, J. S. (1990). *The city as text: The politics of landscape interpretation in the Kandyan kingdom*. Cambridge, UK: Cambridge University Press.

Escobar, A. (1995). *Encountering development: The making and unmaking of the third world*. Princeton, NJ: Princeton University Press.

Escobar, A. (1998). Whose knowledge, whose nature?: Biodiversity, conservation, and the political ecology of social movements. *Journal of Political Ecology, 5*, 53–82.

Florescano, E. (1980). *Análisis histórico de las sequías en México*. Mexico City: Secretaría de Agricultura y Recursos Hidráulicos.

García, E. (1970). Los climas del estado de Veracruz. *Anales del Instituto de Biología, Serie Botánica, 41*, 3–42.

Gómez-Pompa, A. (1973). Ecology of the vegetation of Veracruz. In A. Graham (Ed.), *Vegetation and vegetational history of northern Latin America* (pp. 73–148). Amsterdam: Elsevier.

Harley, J. B. (1992). Rereading the maps of the Columbian encounter. *Annals of the Association of American Geographers, 82*, 522–542.

Latour, B. (1993). *We have never been modern*. Cambridge, MA: Harvard University Press.

Lockhart, J., & Otte, E. (Eds.). (1976). *Letters and people of the Spanish Indies, sixteenth century*. Cambridge, UK: Cambridge University Press.

Paso y Troncoso, F. del. (Ed.). (1905). *Papeles de Nueva España*. Madrid: Sucesores de Rivadeneyra.

Peet, R., & Watts, M. (Eds.). (1996). *Liberation ecologies: Environment, development, social movements*. New York: Routledge.

Pérez de Rivas, A. (1896). *Crónica y historia religiosa de la provincia de la Compañía de Jesús de México*. Mexico City: Sagrado Corazon de Jesús.

Rojas Rabiela, T. (1988). *Las siembras de ayer*. Mexico City: Secretaría de Educación Pública.

Rosenberg, N. (1982). Natural resource limits and the future of economic progress. In G. A. Almond, M. Chodorow, & R. H. Pearce (Eds.), *Progress and its discontents* (pp. 301–318). Berkeley and Los Angeles: University of California Press.

Sachs, W. (Ed.). (1992). *The development dictionary: A guide to knowledge as power*. London: Zed Books.

Sartorius, C. C. (1961). *Mexico about 1850*. Stuttgart, Germany: F. A. Brockhaus.

Sauer, C. O. (1956). Summary remarks: Retrospect. In W. L. Thomas (Ed.), *Man's role in changing the face of the Earth* (pp. 1131–1135). Chicago: University of Chicago Press.

Sluyter, A. (2002). *Colonialism and landscape: Postcolonial theory and applications.* Lanham, MD: Rowman & Littlefield.

Sluyter, A. (2003). *Scientific geography and the modern nature/society and West/Rests dichotomies.* Working paper available in manuscript from the author, Department of Geography and Anthropology, Louisiana State University, 227 Howe-Russell Geoscience Complex, Baton Rouge, LA 70803-4105.

Souza, A. R. de. (1986). To have and have not: Colonialism and core–periphery relations. *Focus, 36*(3), 14–19.

Stott, P., & Sullivan, S. (Eds.). (2000). *Political ecology: Science, myth and power.* London: Arnold.

Turner, B. L. II, & Butzer, K. W. (1992). The Columbian encounter and land-use change. *Environment, 34*(8), 16–20, 37–44.

United Nations. (1993). *Agenda 21: The United Nations programme of action from Rio.* New York: Author.

Wolf, E. R. (1982). *Europe and the people without history.* Berkeley and Los Angeles: University of California Press.

World Watch Institute. (2002). *State of the world 2002: Special world summit edition.* New York: Norton.

Zimmerer, K. S. (1996). *Changing fortunes: Biodiversity and peasant livelihood in the Peruvian Andes.* Berkeley and Los Angeles: University of California Press.

CHAPTER 12

The Production of Nature

Colonial Recasting of the African Landscape in Serengeti National Park

Roderick P. Neumann

This chapter analyzes the history of the establishment of what was meant to be the first national park in British-ruled Africa: Serengeti. It illustrates how European preservationists' efforts to recast the African landscape into their ideal of a wilderness continent resulted in a protracted struggle over land and resource rights in the park. The analysis of this struggle in colonial Tanganyika (now Tanzania) reveals a process of nature *production* rather than nature *preservation* in the establishment of Serengeti National Park. The idea of nature as a pristine empty African wilderness was largely mythical and was made concrete only by relocating thousands of Africans and denying millennia of human agency in shaping the landscape. The history of Serengeti's creation demonstrates that while processes of development and processes preservation are often seen to be in opposition and clearly produce very different landscapes, they were in fact linked by a shared ideology: the assertion of European control over society and nature in Africa.

The study is primarily based on examination of archival documents of the British colonial administration of Tanganyika. Valuable as they are for re-

Adapted by Roderick P. Neumann from the article "Ways of Seeing Africa: Colonial Recasting of African Society and Landscape in Serengeti National Park," published in *Ecumene*, 2(2), 149–169 (1995). Adapted by permission of Arnold Publishers. The adapting author is solely responsible for all changes in substance, context, and emphasis.

constructing the case history of Serengeti National Park, documents written by colonial administrators predictably contain biases and shortcomings. It is therefore important that they be supplemented with other sources (Watts, 1983). The archival research featured here was complemented by on-site research with displaced groups organized to reclaim their lost land and resource rights in the national park (see Neumann, 1995). These latter sources are important for unveiling alternative landscape readings and environmental histories that contest the popular and prevailing narratives that equate African protected areas with timeless wilderness.

The case history is framed by theories of the social production of nature as they relate to the Western concept of landscape. The theoretical development of the idea of nature as social product is most closely identified with Smith's (1984) *Uneven Development*. To overcome the society–nature dualism of positivist approaches to external nature, Smith proposed two concepts: "first nature," that which is unaltered by humans; and "second nature," that which is incorporated into the institutions—the market, the state, money—that have developed to regulate commodity exchange. Under capitalism, first nature disappears as it enters the realm of exchange value and becomes just another commodity, second nature. This is the case even of supposedly pristine nature in national parks, which has been commodified as "neatly packaged cultural experiences of environment" (Smith, 1984, p. 57).

Denis Cosgrove (1984) developed a theory of landscape through arguments that parallel Smith's. Tracing historically the changing ideas of landscape in England, he argued that as land and labor became increasingly oriented toward production for exchange, they became less capable of portraying idealized (read, naturalized) social relations. The landscape aesthetic in painting and literature refocused on countryside settings where evidence of human labor was erased or hidden. The resulting effect was the creation, materially and symbolically, of two distinct landscapes: one of production ruled by rationality and profit and one of consumption where recreation and contemplation prevail (see also Frykman & Lofgren, 1987; Williams, 1973). By the end of the 19th century, the idea of landscape as an object of aesthetic appreciation became nearly synonymous with the idea of nature.

Through colonialism, the "landscape way of seeing" (Cosgrove, 1984) was transported to East Africa. In colonial Tanganyika, British efforts to preserve nature in national parks coincided with efforts to intensify agricultural production in the territory, that is, to develop nature rather than to preserve it. The introduction of these spatially distinct ideologies, consumption and production, preservation and development, had major ramifications for the transformation of African land rights and land use practices. The counterpart of this spatial dualism is stereotypical British colonial views of African culture, which can be simplified into two ideological currents. On the one hand, the British romanticized precontact African society through such ideas as the moral innocence of Africans, a respect for African bush skills, and a generalized notion of the noble savage (Curtin, 1964; Grove, 1995). But on the other hand, Africans were seen as untapped labor who through development and modernization could become efficient producers within the sphere of the British colonial economy.

National park advocates used the traditional and modern stereotypes simultaneously in pushing their agenda. That is, they argued that confining wild animals in parks would actually promote development elsewhere by reducing conflicts between wildlife and African agricultural. Within parks, preservationists argued that resident Africans could be allowed to stay if they continued to live "traditionally." The fact that human labor was denied in landscapes of consumption as conceptualized by European preservationists meant a significant alteration of land use and land access for Africans. At Serengeti, British notions of who was or was not Maasai, who had customary rights of occupation, and what "traditional" land use practices would be allowed there framed the political conflict that ultimately led to the forced evacuation of the park.

IMAGINING WILD AFRICA

Proposals to preserve nature in national parks in East Africa began in the early years of the 20th century. In Tanganyika, local game officials first proposed establishing a system of national parks in the 1920s, although the strongest push for the idea originated outside the colony. In 1931, the London-based conservation organization the Society for the Preservation of the Fauna of the Empire (SPFE) spearheaded the cause, sending their representative Major Richard Hingston to Tanganyika to investigate the needs and potential for developing a nature preservation program. The report that resulted stressed that "the keystone . . . is the recommendation regarding the formation of a number of national parks without delay."[1] Hingston's report provided the basis for the SPFE's efforts to forge an international agreement. The Convention for the Protection of the Flora and Fauna of Africa was held in London in 1933, resulting in an international agreement that closely followed the SPFE's proposals. A section in the London Convention obligated all signatories to investigate the possibilities for national parks in their respective colonies.

It is clear from Hingston's report and the London Convention that the prime interest of European preservationists was to reserve wild nature in Africa in a system of national parks. The nature preservationists were driven in part by a fear of losing "Eden," a fear that their vision of Africa as primeval wilderness was fading before their eyes (Anderson & Grove, 1987). The sentiments of nostalgia and loss that drive this fear have been common in European preservationists' thinking and have been particularly strong in the settler colonies of East and southern Africa. National parks are meant to represent primordial, undisturbed, and unchanging nature in Africa. The definition, designation, and regulation of national parks were, to a large degree, concerned with making ecological reality conform to this imagined African landscape.

National park advocates had to confront the fact that there were established African societies living within the proposed boundaries of national parks. If European preservationists found Africans to be living in what they deemed a "natural state," then they could be protected within parks as another native species. If Africans deviated from the ideal of the "natural," their presence came into conflict with the preservationists' vision. The SPFE made clear that humans were not welcome in parks, but pointed out that a "native" presence in the parks might be tolerated, as in the case of Parc National Albert (Belgian Congo) where "the Pygmies are rightly *regarded as part of the fauna*, and they are therefore left undisturbed" (emphasis added).[2]

Preservationists crudely constructed ideas of "traditional culture" and society–nature relationships in Africa. That is, their ideas were formed not from detailed ethnographic knowledge, but from long-standing stereotypes about African race and culture, particularly the notion that some Africans were living in a natural state. In the extreme, preservationists considered Africans as morally and socially equivalent to wildlife species in need of protection. These ideas can be traced to the historical process of colonial conquest, where the line between hunting animals and making war on Africans was often blurred (MacKenzie, 1987). Once conquest was complete, these views persisted in less pernicious form and were incorporated into the discourse of nature preservation.

It was the European preservationists' prerogative, moreover, to determine the precise character of "primitive culture." That is, just as there was a particular European landscape ideal of unspoiled nature which Africa represented, there was an interrelated concept of primitive human society that Africans represented. Both had more to do with European myths and desires than with African reality. Those Africans whose behavior did not fit with British preconceptions of "primitive man" could not be allowed to remain in the national parks, the symbol of primeval Africa, regardless of their claims to customary land rights. National parks were at once symbolic representations of the European vision of Africa and a demonstration of the colonial state's power to control access to land and natural resources.

HARNESSING THE "VAST UNDEVELOPED CONTINENT"

The disruption of African land use practices resulting from nature preservation programs was a point of political friction not only between Africans and Europeans, but also within European colonial society. British colonial thinking was far from monolithic, and there was strong opposition to the preservationists' ideas. Throughout the British colonial period, administrative officers continually criticized the national park proposals for interfering with African land and resource rights. From their perspective, for example, hunting by African residents inside the proposed parks was acceptable, since "when such natives have enjoyed customary rights of hunting there is no reason or justification for depriving them of these rights."[3] A. E. Kitching, a district officer and later provincial commissioner, exemplified this position. To Major Hingston's 1931 report he responded:

> The Hingston recommendations . . . pay no regard to native interests. They involve the alienation in perpetuity of thousands of square miles of the land of the Territory . . . to "create the finest nature park in the Empire." The recommendations appear to me to be so wrong in principle as to make any detailed examination unnecessary.

Kitching's response reflected the thinking of a large contingent of colonial officials at all levels who were more concerned with the economic and social development of Africa than with preservation of its wildlife. Beginning in the 1930s (but only making significant advancement after World War II) development advocates took an ideological position that ran counter to the preservationists' visions of primitive Africa. While the SPFE and others were working to preserve nature, those trying to develop the productive potential of the territory were working hard to eradicate it. In many areas of East and southern Africa, wilderness was seen to be increasing since the early 20th century as an advancing tsetse fly front (and the sleeping sickness that accompanied it) forced the evacuation of thousands of square kilometers (Neumann, 2002).

In fact, the Tanganyika Game Preservation Department was itself originally organized principally to control rather than to protect wildlife. Charles Swynnerton was recruited in 1919 to be its first director based on his experience with tsetse fly eradication efforts, which in those days were practically synonymous with the eradication of wildlife habitat.[5] Swynnerton saw the elimination of tsetse as *the* key to the commercial exploitation of Africa:

> Tsetse are actually holding up . . . the development of a continent . . . holding up also vast grazing areas which, if they could be stocked, would . . . provide homes not merely for Natives, but for settlers . . . The vast undeveloped Continent of Africa can supply England's need of raw materials, but she can never be developed adequately while the tsetse remains in possession.

The Tanganyika Game Ordinance of 1921 (drafted by Swynnerton) included a schedule of game reserves within which wildlife would receive varying degrees of protection. Whether engaged in habitat alteration or preservation, the ideology underlying the agenda of the Game Preservation Department was control over nature.

British agriculture and livestock specialists were critical of African practices, advocating direct state intervention and the application of scientific principles in land use. As the director of the Department of Veterinary Science and Animal Husbandry put it, "We are trying to stop them [Africans] from being their own worst enemies" (Tanganyika Territory, 1936, p. 31). He argued that "traditional native husbandry" was essentially inefficient and destructive, a situation that could only be rectified by "some form of government intervention" (Tanganyika Territory, 1935, p. 147).

Efforts to modernize agricultural production and harness African labor intensified in the 1940s. The postwar period in East Africa was marked by a "second colonial occupation" (Low & Lonsdale, 1976, p. 12), characterized by new education programs for Africans, infrastructural development, and cash-crop expansion (Iliffe, 1979). These programs and policies required increased intervention to raise rural production and hence a deeper penetration of the colonial state into nearly all aspects of African society (Beinart, 1984). In other words, postwar development would channel African labor into greater and more efficient production through the collateral transformation of nature. This agenda was institutionalized in 1946 when the government established the Development Commission, which issued the first 10-year development and welfare plan a year later.[7] Outside the scope of the development plan but directly related to its broad goals were a number of agricultural schemes intended to introduce new export crops and a new social organization of production. Some of these will be examined later in the discussion of Serengeti National Park's establishment.

DEFINING RIGHTS, PRODUCING NATURE

Following the signing of the 1933 London Convention, the Colonial Office in England increasingly pressured the government in Tanganyika to comply with its terms. Specifically, they were urged to investigate the possibilities for a system of national parks. Tanganyikan game officials concentrated their attention on the Serengeti region near the border with Kenya, which the government had already placed under protection as a game reserve under the 1921 Game Ordinance. The level of protection, however, was insufficient for nature preservationists, and the SPFE soon focused on Serengeti in its efforts to establish the first national park in British-ruled Africa.

As the pressure for park establishment increased, members of the colonial administration in Tanganyika voiced their concerns over the political problems that could arise with resident Africans and over the impact that preservation would have on agricultural production. For example, the initial proposal for Serengeti National Park in Hingston's report would have placed 70,000 Africans under the authority of the park in one district alone. Some colonial officials argued that African land rights were protected under the League of Nations Trusteeship doctrine and that any restrictions on these rights would adversely affect their livelihood. Far and away the most prevalent theme of the comments of Tanganyikan officials was a concern that the parks would interfere with local customary rights to grazing, hunting, and minor forest products.[8] Consequently, acting governor D. J. Jardine wrote to the secretary of state for the colonies asking that a clause be added "to the effect that the protection of vegetation in national parks does not interfere with the rights at present enjoyed by the native inhabitants to pasture or to forest produce."[9] The secretary's eventual reply points out that nothing in the national park definition makes "native" habitation inconsistent, "provided that they are controlled by Park authorities."[10]

The government in Tanganyika eventually drafted a new game ordinance that included a clause declaring Serengeti a national park. The governor established a special committee to review the bill. The committee reiterated government concerns over indigenous rights in the proposed park. The committee recommended "that the requirements of the National Park not be allowed to interfere with existing grazing or water rights."[11] Thus, the residents of the area were allowed to remain and Serengeti was declared a "national park" in the revised 1940 Game Ordinance. This level of legal protection and administrative control was still unsatisfactory to the SPFE and associated preservationists, however, and they lobbied hard for a distinct national park ordinance and an independent board of trustees.

Ultimately, Serengeti National Park took on great symbolic significance for preservationists. This fact helps to explain the SPFE's unceasing lobbying efforts. It came to embody for the society much of what it wished to accomplish for nature preservation in Africa. The society stressed to the government in London

> that the first National Park to be established in a Mandated Territory of British Africa should be modeled as closely as possible on the provisions of the Convention. The Serengeti National Park is likely to serve as a model for others in the British African Colonies and elsewhere and it is on this account that the Committee address the Secretary of State while its constitution is as yet undetermined.

As the prototype national park, then, it was meant to establish in law the preservationists' vision of wild Africa.

The result of SPFE's efforts was the passage of the 1948 National Park Ordinance, which created strict legal protection for Serengeti National Park and established an autonomous governing body, the Serengeti National Park Board of Trustees, to oversee its administration. Significantly, the ordinance explicitly permitted the unhindered movement of people "whose place of birth or ordinary residence is within the park."[13] Still, this left open the difficult question of who did or did not have a legal right to be in the park. The proposed park boundaries were immediately disputed by Africans living there and consequently were not finalized until 1951.[14] Though the secretary of the new Serengeti National Park Board of Trustees reassured the government that "the rights of the Masai . . . to occupy and graze stock in the Park are unaffected by the Ordinance,"[15] less than a week later the new park warden wrote that the trading post and Maasai cattle market in Ngorongoro must be removed because they "interfered with the amenities of the park."[16] It soon became clear that the ordinance constituted a formula for unrest.

ENCLOSURE AND EVICTION

As it happened, Maasai pastoralists had the most to lose in terms of the total area of land that had passed to national park control. They soon initiated acts of protest against the park authority. By the early 1950s, the government had to set up a special administrative post in the park because the "Masai were openly defying the Park laws, and the political situation had consequently become explosive and a magnet for agitators."[17] A further government response to the unrest was to write down in a "bill of rights" precisely what the Maasai who were living in the park could expect. The Maasai did not participate in drawing up the bill of rights and were only allowed to see it after it had been completed.[18] In this way the park officials kept within the written law but violated its spirit. That is, the 1948 National Parks Ordinance explicitly protected customary rights, but only government officials decided what those rights were and who would be entitled to them. Tradition had to be invented (Ranger, 1983) to meet the needs of colonial park authorities. In a case, for instance, where a critical grazing area overlapped with important wildlife habitat, it was reasoned that the Maasai "presence may not be in accordance with strict tribal custom" (Tanganyika Territory, 1957, p. 27). It was the park administration that determined the extent of the customary land of the Serengeti Maasai, restricting their movements to such a small area that the provincial commissioner was moved to call their decision "a complete breach of faith with the Masai."[19]

The question arises, then, upon what historical and ethnographic knowledge was the bill of rights based? Like the so-called Bushmen of Namibia, the Maasai have held a powerful grip on the imagination of European explorers, soldiers, and administrators (Gordon, 1992; Collett, 1987). Early stereotypes from initial contact and conquest filtered into the consciousness of the colonial administrators who followed, crystallizing into a view that "equated nomadic and semi-nomadic pastoralism with a primitive and undeveloped form of social order" (Collett, 1987, p. 138). Based on this conceptualization, the bill of rights written down in Serengeti was essentially a blueprint for an imagined primitive culture. It specified what the basic necessities of material life would be and how they would be acquired. The bill was a reflection of preservationists' desires for a natural relationship between Africans and the land, a statement of their vision of the Maasai as closer to nature than to civilization.

Fulfillment of the European vision of primitive Africans living "amicably amongst the game"[20] meant ignoring economic development and cultural change within resident communities. Under their bill of rights the weapons that the Serengeti Maasai were allowed to carry were restricted to "spears, swords, clubs, bows and arrows."[21] Those Maasai who were allowed to stay were placed under strict control to assure that they remained "primitive." In defining homestead building codes in the Maasai "bill of rights," the chairman of the National Park Board of Trustees

> explained the reasons why they wished the word "traditional" to be inserted in the draft definition in order that the Masai living in the Park should retain their present primitive status. He and the D.N.P. felt that if the Masai changed their habits and wished to build other types of housing, they should do so outside the area of the Park which is to be *reserved as a natural habitat both for game and human beings in their primitive state* [emphasis added].

In the mind of the preservationist, the Maasai in the park were a colonial possession and could be preserved "as part of our fauna."[23]

Like the "Pygmies" of Parc National Albert, the Maasai were imagined to be living more or less harmoniously with nature because they were nomadic, did not hunt, and generally did not cultivate. When Africans did not live up to European stereotypes, attempts were made to make them conform. In the context of Serengeti, these attempts generated more conflict. For instance, some Maasai did in fact cultivate,[24] though the national parks director tried to explain the presence of cultivators in Ngorongoro Crater as a result of the Maasai having become "much adulterated with extra-tribal blood."[25] Based on British interpretations of African culture, the legal logic by which these cultivators could be evicted ran as follows: since we know that Maasai did not traditionally cultivate, any cultivators in the crater must be non-Maasai, and since no non-Maasai may live in Maasailand without a permit from the Native Authority, they are therefore without legal rights.[26]

Other land use practices that have been historically important to pastoralism in East Africa were unacceptable as well, particularly the use of fire to manipulate vegetative growth. Fire was a critical element in Maasai pasture management, eradicating disease-bearing ticks and maintaining grasslands (Tanganyika Territory, 1929, p. 8; 1933, p. 81). Infested pastures would be temporarily abandoned and burned until the threat of disease had been removed. Fires in the highlands surrounding Ngorongoro Crater and elsewhere would open up forest glades of high-quality forage. The 1940 Game Ordinance, however, outlawed this practice within Serengeti National Park, though the park administration was unsuccessful in its efforts to stop it. Fire thus became a point of struggle between park officers and resident Maasai. The preservationists were, in sum, inventing Maasai tradition as the need arose, banning some practices and freezing others from change in an effort to mold the Maasai to their national park ideal. This effort coincided with and was inseparable from their efforts to re-create the imagined landscape of Serengeti National Park.

The Maasai were not the only residents in the park. The Ndorobo and the Sukuma, among others, hunted and cultivated. Though nothing in the law prohibited the practice, park officials had all along planned to evict cultivators. Soon after the park was gazetted they began plans to amend the ordinance to explicitly forbid cultivation.[27] The 1948 National Park Ordinance contained an inherent contradiction. The "saving clause" read,

> Nothing in this Ordinance contained shall affect . . . the rights of any person in or over any land acquired before the commencement of this Ordinance.

Yet the law also gave the trustees power to make regulations concerning hunting and a variety of land use practices, which would almost certainly interfere with existing rights. The ambiguity of the law had the effect of fanning the flames of discontent as administrative officers would tell park residents that their rights were fully protected, while park authorities would try to enforce regulations to restrict their activities. The government's "solution" was to pass an amended law in 1954, which removed the contradiction by revoking the saving clause, and replacing it with one that expressly denied any right of occupants to cultivate and gave the governor extraordinary powers to prohibit any other activities deemed undesirable.

The actions of the board of trustees and the park staff contrasted starkly with the government's development schemes operating outside the boundaries of the park. In an adjacent district, a development project was initiated in the late 1940s to, among other things, introduce "more intensive methods of agriculture" and to clear fly-infested bush for relocations of households and livestock (Moffett, 1955, p. 375). Maasailand had the largest of the local schemes in terms of area, encompassing 23,000 square miles. The scheme's main objective was to "improve Masailand as a ranching country" and "induce a more stable economy" (Moffett, 1955, p. 531). Scheme managers set about clearing thousands of acres of wildlife habitat with bulldozers with intentions to build roads, schools, and dispensaries. These schemes on the edge of Serengeti further sharpened the distinction and widened the symbolic and ecological gap between the landscapes of production and consumption.

In Serengeti, meanwhile, park officials began making plans to evict cultivators living within the park.[29] Acknowledging that the National Park Ordinance legally protected cultivators' rights, officials proceeded anyway, though worrying that "their eviction is not going to pass unnoticed among local agitators and it is therefore important that it should receive legal sanction."[30] The disputes between park residents and the park administration evolved within the context of a rising tide of African nationalism across East Africa fueled by the politics of land. In 1951, the year that the Tanganyika government finally set Serengeti's boundaries, the British colonial government in Kenya declared the Mau Mau Emergency, centered just 3 hours' drive from the park headquarters. The fighting there between the African Land and Freedom Armies and the colonial government continued throughout the period of greatest unrest in Serengeti.

As East Africa threatened to explode over the issue of African land rights, the preservationists were advocating, and gaining government support for, the dislocation of hundreds of families to create what was popularly viewed as a playground for white tourists. Rather than retreating in the face of growing nationalist sentiments surrounding African land rights, nature preservationists pushed the issue. As unrest among park residents grew, the preservationists' position hardened and became less ambiguous over the issue of human occupation in a national park. "The interests of fauna and flora must come first," a park manager wrote, "those of man and belongings being of secondary importance. Humans and a National Park can not exist together."[31]

Something clearly had to give. In April 1956 the government published a report on the problems, recommending that Serengeti be reconstituted so that nature preservation and human interests be spatially segregated (Tanganyika Territory, 1956). Preservationists objected to the recommendations because the suggested area was too limited for wildlife migrations and they compelled the government to appoint a committee of inquiry to revisit the situation.

Many of the concerns about the future of the park were now stated in the discourse of the second colonial occupation, with its influx of technocrats trained in applied fields such as agronomy and soil science (Beinart, 1984). By now the SPFE had changed its name to the was now the Fauna Preservation Society (FPS). Their agents in the field were no longer big game hunters and former soldiers, but scientists. The FPS sent Professor W. H. Pearsall, a biologist, to Serengeti to produce their report, which ultimately provided the basis of the committee's findings. The committee called for a "more scientific approach" (Tanganyika Territory, 1957, p. 16) as they sought to ensure that the park would function as a "viable ecological unit" (p. 23). Within this scientific approach, the idea that humans and nature must be spatially separated was now referred to as a "principle" (Tanganyika Territory, 1957, p. 15).

At the heart of the committee's recommendations was an endorsement of the principle that human rights should be excluded in any national park. The committee recommended that the national park should be reconstituted in the Western Serengeti, and that the Ngorongoro Crater sector be excised from the park and managed as a special conservation unit where Maasai pastoralists would be allowed to stay. Summing up the resulting National Park Ordinance (Amended) of 1959, the chairman of the board of trustees wrote:

> Under this ordinance the Tanganyika National Parks become for the first time areas where all human rights must be excluded thus eliminating the biggest problem of the Trustees and the Parks in the past. (Tanganyika National Parks, 1960)

DISCUSSION

There are three points that I would like to draw from this case history. First, the establishment of national parks in colonial Tanganyika was as much a process of nature *production* as of nature *preservation*. The incorporation of Tanganyika into the global economy meant a pattern of landscape partitioning: here intensive agriculture and ranching, there nature preserved from the forces of capitalist production. Nature was produced in national parks based on a preconceived, culturally constituted vision of Africa as primeval wilderness. Initially, the vision could include the people who claimed customary rights of occupation and use, because they could be considered to be part of primeval nature. Within a Eurocentric evolutionary view of culture, hunters and gatherers and nomadic pastoralists were considered to be living more off the fruits of nature than their own labor and would not therefore necessarily disrupt the landscape aesthetic. Ultimately, however, the myth of the Maasai as "natural" humans could not be sustained as preservationists were increasingly confronted with the evidence of their labor and agency. Thus nature, as represented in national parks, was produced by removing the people who, ironically enough, had influenced the ecology of the Serengeti through thousands of years of human agency (Collett, 1987; Homewood & Rodgers, 1987).

Second, the colonial government's efforts at intensifying agricultural and livestock production through development schemes near Serengeti National Park created contradictions and sharp distinctions between the landscapes of production and those of consumption. Yet despite the contrasting effects on African society and landscape, preservation and development projects sprang from the same ideological roots. There existed in each program an unfaltering belief in the superiority of European culture and science and a contemptuous view of African land use practices. In both projects, these beliefs served to justify the violent shift of control over land and resources from local lineages to the state. Both were about controlling nature, one for aesthetic purposes, and the other for intensifying production. Borrowing from Raymond Williams's analysis of the transition to capitalist property relations in England, development and preservation "are related parts of the same process—superficially opposed in taste but only because in the one case the land is being organized for production . . . while in the other case it is being organized for consumption—the view" (Williams, 1973, p. 124).

Ultimately, in agrarian societies, control over nature equates to control over people. In each case, African production practices had to be removed, reoriented or reorganized depending on the needs and desires of the colonizers. The two colonial programs—preservation and development—thus introduced a new spatial separation of culture and nature, production and consumption, thereby transforming the African landscape and severing historical human–land relationships and customary control over land and resources.

Finally, the history of Serengeti National Park can be read as a cautionary tale for contemporary efforts to establish and maintain parks and protected areas in Africa. Notions of static African societies living in concert with nature persist in present-day protected area proposals. Namibia is a case in point. In the 1980s, the government planned to allow the Ju/wasi people to remain inside a proposed game park if they hunted only with bows and arrows (Volkman, 1986). Economic development, such as ranching, was prohibited because it violated the imposed definition of "traditional" culture. The Ju/wasi would become, in essence, another tourist attraction—this in spite of the fact that far from being "primitive" or "traditional," they had been incorporated into the world economy in various ways for over a century (Gordon, 1992).

Such protected area designs beg the question of inevitable cultural change. Proponents of these culture/nature reserves often gloss over the surrender of political power and loss of control over daily life that is required of would-be residents. Proposals such as the Ju/wasi game park also make clear that Western preservationists are still captivated by a way of seeing traditional African society as living in aesthetic harmony with the landscapes of nature.

NOTES

1. Hingston, "Report on a Mission to East Africa," TNA Secretariat File 19038.
2. Report of the delegates of the International Congress for the Protection of Nature, Paris, June 1931, to His Majesty's Government; and "Note on the Convention," anon., n.d., TNA Secretariat File 12005.
3. Under Secretary of State to SPFE, 2/10/39, TNA Secretariat File 12005.
4. Comments on Major Hingston's Report on a Mission to East Africa, 9/4/31, TNA Secretariat File 12005. Comments on Major Hingston's Report on a Mission to East Africa, 9/4/31, TNA Secretariat File 12005.
5. Roger Swynnerton, "Forward" to "Tsetse Research and Reclamation in Tanganyika

Territory and the Role of C.F.M. Swynnerton, C.M.G.," Rhodes House (henceforth RH), MSS.Afr.s.1987.

6. C.F.M. Swynnerton, "The Problem of the Tsetse," lecture delivered to the Royal Colonial Institute, March 1925, April 16, 1925, RH, MSS.Afr.s.1987.

7. Report to the Trusteeship Council of the United Nations on the Administration of Tanganyika, 1947, London.

8. Summary of Observations on the Report of the Preparatory Committee, n.d., TNA Secretariat File 12005.

9. Acting Governor Jardine to Secretary of State Cunliffe-Lister, 1/8/33, TNA Secretariat File 12005.

10. Secretary of State to Governor MacMichael, 17/3/34, TNA Secretariat File 12005.

11. Report of the Special Committee Appointed to Examine the Game Bill, 1940, 16/4/40, TNA Secretariat File 27273.

12. SPFE Secretary to Under Secretary of State, 29/8/39, TNA Secretariat File 12005.

13. Tanganyika Territory National Parks Ordinance, 1948.

14. Memorandum No. 82 for Executive Council, 22/8/50, TNA Secretariat File 34819.

15. P. Bleackley, Secretary, Serengeti National Park Board of Trustees to Member for Local Government, Dar Es Salaam, 18/10/51, TNA Secretariat File 10496.

16. Minutes of the second meeting of the Serengeti National Park Board of Trustees, 23/10/51, TNA Secretariat File 10496.

17. District Commissioner Masai/Monduli to District Officer Ngorongoro, 5/3/55, TNA Arusha Regional File G1/6, Accession No. 69.

18. Acting Provincial Commissioner, Northern Province, to Member for Local Government, 20/5/55, TNA Arusha Regional File G1/6, Accession No. 69.

19. Provincial Commissioner, Northern Province, to Member for Local Government, 28/2/55, TNA Arusha Regional File G1/6, Accession No. 69.

20. Major R. Hingston, *Report on a Mission to East Africa*, TNA Secretariat File 19038.

21. Provincial Commissioner, Northern Province, to Member for Local Government, 19/1/55, TNA Arusha Regional File G1/6, Accession No. 69.

22. Notes on a meeting between the Chairman of the National Park Board of Trustees, the Director of National Parks, and the Provincial Commissioner, Northern Province, 28/3/55, TNA Arusha Regional File G1/6, Accession No. 69.

23. Barclay Leechman, Chairman of the Serengeti National Park Board of Management, in the minutes of the Serengeti National Park Board of Management meeting, 23/7/53, TNA Secretariat File 40851.

24. Acting Provincial Commissioner, Northern Province, to Director of National Parks, 8/6/55, TNA Arusha Regional File G1/6, Accession No. 69. The provincial commissioner pointed out that a recent census had determined that 82 out of 216 families cultivating in the Crater were Maasai.

25. Tanganyika National Parks Director Molloy to Provincial Commissioner, Northern Province, Report on Human Inhabitants, Serengeti National Park, 8/6/55, TNA Arusha Regional File G1/6, Accession No. 69.

26. Notes on an informal discussion held at the Ngorongoro Rest Camp among members of the SNP Boards of Trustees and Management, TNA Arusha Regional File T3/2, Accession No. 69.

27. Confidential letter, District Commissioner, Masai/Monduli, to Provincial Com-

missioner, Northern Province, 23/6/52, TNA Arusha Regional File T3/2, Accession No. 69.

28. Tanganyika Territory, National Parks Ordinance, 1948.

29. Notes on an informal discussion held at the Ngorongoro Rest Camp among members of the SNP Boards of Trustees and Management, TNA Arusha Regional File T3/2, Accession No. 69. The notes state that the park administration "would be both willing and able to evict all non-Masai from the park within a year or two."

30. Confidential letter, District Commissioner Masai/Monduli to Provincial Commissioner, Northern Province, 23/6/52, TNA Arusha Regional File T3/2, Accession No. 69.

31. Wilkins, SNP Board of Management, to SNP Board of Trustees, 16/2/54, TNA Secretariat File 10496.

REFERENCES

Anderson, D., & Grove, R. (1987). Introduction: The scramble for Eden: Past, present and future in African conservation. In D. Anderson & R. Grove (Eds.), *Conservation in Africa: People, policies and practices* (pp.1–12) Cambridge, UK: Cambridge University Press.

Beinart, W. (1984). Soil erosion, conservationism and ideas about development: A southern African exploration, 1900–1960. *Journal of Southern African Studies, 11*(1), 52–83.

Collett, D. (1987). Pastoralists and wildlife: Image and reality in Kenya Maasailand. In D. Anderson & R. Grove (Eds.), *Conservation in Africa: People, policies and practices* (pp. 129–148). Cambridge, UK: Cambridge University Press.

Cosgrove, D. (1984). *Social formation and symbolic landscape.* London: Croom Helm.

Curtin, P. (1964). *The image of Africa: British ideas and action, 1780–1850.* Madison: University of Wisconsin Press.

Frykman, J., & Lofgren, O. (1987). *Culture builders: A historical anthropology of middle-class life.* New Brunswick, NJ: Rutgers University Press.

Gordon, R. (1992). *The Bushman myth: The making of a Namibian underclass.* Boulder, CO: Westview Press.

Grove, R. (1995). *Green imperialism.* Cambridge, UK: Cambridge University Press.

Homewood, K., & Rodgers, W. (1987). Pastoralism, conservation and the overgrazing controversy. In D. Anderson & R. Grove (Eds.), *Conservation in Africa: People, policies and practices* (pp. 111–128). Cambridge, UK: Cambridge University Press.

Iliffe, J. (1979). *A modern history of Tanganyika.* Cambridge, UK: Cambridge University Press.

Low, D., & Lonsdale, J. (1976). Introduction. In D. Low & A. Smith (Eds.), *The Oxford history of East Africa* (pp. 1–64). Oxford, UK: University of Oxford Press.

MacKenzie, J. (1987). Chivalry, social Darwinism and ritualized killing: The hunting ethos in Central Africa up to 1914. In D. Anderson & R. Grove (Eds.), *Conservation in Africa: People, policies and practices* (pp. 41–62). Cambridge, UK: Cambridge University Press.

Moffett, J. P. (Ed.). (1955). *Tanganyika: A review of its resources and their development*. Dar Es Salaam: Government Printers.

Neumann, R. P. (1995). Local challenges to global agendas: Conservation, economic liberalization, and the pastoralists' rights movement in Tanzania. *Antipode, 27*(4), 363–382.

Neumann, R. P. (2002). The postwar conservation boom in British colonial Africa. *Environmental History, 7*(1), 22–47.

Ranger, T. (1983). The invention of tradition in colonial Africa. In E. Hobsbawm & T. Ranger (Eds.), *The invention of tradition* (pp. 211–262). Cambridge, UK: Cambridge University Press.

Smith, N. (1984). *Uneven development: Nature, capital and the production of space*. Oxford, UK: Basil Blackwell.

Tanganyika National Parks. (1960). *Reports and accounts of the Board of Trustees July 1959 to June 1960*. Arusha: East African Printers Tanganyika.

Tanganyika Territory. (1929). *Annual report of the Department of Veterinary Science and Animal Husbandry, 1929*. Dar Es Salaam: Government Printers.

Tanganyika Territory. (1933). *Annual report of the Department of Veterinary Science and Animal Husbandry, 1933*. Dar Es Salaam: Government Printers.

Tanganyika Territory. (1935). *Annual report of the Department of Veterinary Science and Animal Husbandry, 1935*. Dar Es Salaam: Government Printers.

Tanganyika Territory. (1936). *Annual report of the Department of Veterinary Science and Animal Husbandry, 1936*. Dar Es Salaam: Government Printers.

Tanganyika Territory. (1956). *The Serengeti National Park* (Tanganyika Sessional Paper No. 1 of 1956). Dar Es Salaam: Government Printers.

Tanganyika Territory. (1957). *Report of the Serengeti Committee of Enquiry, 1957*. Dar Es Salaam: Government Printers.

Tanganyika Territory. (1958). *Proposals for reconstituting the Serengeti National Park* (Tanganyika Sessional Paper No. 5 of 1958). Dar Es Salaam: Government Printers.

Volkman, T. A. (1986). The hunter–gatherer myth in southern Africa. *Cultural Survival Quarterly, 10*(2), 25–32.

Watts, M. (1983). *Silent violence: Food, famine, and peasantry in northern Nigeria*. Berkeley and Los Angeles: University of California Press.

Williams, R. (1973). *The country and the city*. New York: Oxford University Press.

CHAPTER 13

Agroenvironments and Slave Strategies in the Diffusion of Rice Culture to the Americas

Judith Carney

By the mid-1700s a distinct cultivation system, based on rice, shouldered the American and African Atlantic. One locus of rice cultivation extended inland from West Africa's Upper Guinea Coast, another flourished along the coastal plain of South Carolina and Georgia, and a third developed in the corridor between Brazil's Northeast and the eastern Amazon region (Figure 13.1). This tidal rice cultivation system depended upon enslaved African labor. In West Africa farmers planted rice as a subsistence crop on small holdings, with surpluses occasionally marketed, while in South Carolina, Georgia, and Brazil, cultivation depended on a plantation system and West African slaves to produce a crop destined for international markets.

While rice cultivation continues in West Africa today, its demise in South Carolina and Georgia swiftly followed the abolition of slavery. Brazil's experiment in tidal rice cultivation, modeled after that of Carolina, did not withstand its competition in the 19th century. Yet the U.S. South's most lucrative plantation economy continued to inspire nostalgia well into the 20th century when the crop and the princely fortunes it delivered remained no more than a

Adapted by Judith Carney from the article "Landscaoes of Technology Transfer: Rice Cultivation and African Continuities," published in *Technology and Culture, 31*(1), 5–35 (1998). Copyright by the Society for the History of Technology. Adapted by permission of The Johns Hopkins University Press. The adapting author is solely responsible for all changes in substance, context, and emphasis.

FIGURE 13.1. Rice cultivation along the Atlantic Basin, 1760–1860.

vestige of the coastal landscape. Numerous commentaries documented the lifeways of European American planters, their achievements, and their ingenuity in shaping a profitable landscape from malarial swamps (Doar, 1936/1970; Flanders, 1933; Heyward, 1937). These accounts have never presented African Americans as having contributed anything but their unskilled labor. The planter-biased rendition of the origins of American rice cultivation prevailed until 1974 when the historian Peter Wood carefully examined the role of slaves in the Carolina plantation system during the colonial period. His scholarship recast the prevalent view of slaves as mere field hands to one that showed that they contributed agronomic expertise as well as skilled labor to the emergent plantation economy (Wood, 1974). Littlefield (1981) built upon Wood's pathbreaking thesis by discussing the antiquity of African rice-farming practices and by revealing that more than 40% of South Carolina's slaves during the colonial period originated in West Africa's rice cultivation zone.

While this research has resulted in a revised view of the rice plantation economy as a fusion of both European and African cultures, the agency of African slaves in its evolution is still debated. Current formulations question whether planters recruited slaves from West Africa's rice coast to help them develop a crop whose potential they independently discovered, or whether African-born slaves initiated rice planting in South Carolina by teaching planters to grow a preferred food crop. The absence of archival materials that would document a tutorial role for African slaves is not surprising given the paucity of records available in general for the early colonial period, and because racism over time institutionalized white denial of the intellectual capacity of bondsmen. An understanding of the potential role of slaves demands other forms of historical enquiry.

This study is situated in a growing trend in scholarship that integrates detailed ethnographic and ecological investigation, particularly of agroenvironments, with social and environmental history. Such an integration invites the use of multiple and diverse tools, including analysis of archival materials and historical documents (e.g., travelers' narratives, colonial accounts, maps), oral histories, agroecological methods, and ethnographic inquiry (Egan & Howell, 2001; Fairhead & Leach, 1996). The following discussion of West African rice-farming technology and culture is complemented by extensive fieldwork conducted on rice systems in Senegambia by the author (Carney, 2001b), by Olga F. Linares in Senegal (1992), and by Paul Richards in Sierra Leone (1986). Similarities between today's West African rice culture and that of the antebellum U.S. South do not prove in themselves the case for rice technology transfer by African slaves. Field-based geographical studies, however, can substantiate and elaborate upon often-sparse observations in the archival record; provide theoretical frameworks, inspired by political ecology, for understanding past human–environment relations; and produce a richer portrait of past agricultural landscapes.

This chapter combines geographical and historical perspectives to examine the likely contributions of African-born slaves to the colonial rice economy. The approach identifies and describes the principal West African microenvironments planted to rice in the first section as the basis for examining, in the one following, the systems that emerged in South Carolina during the colonial period. While archival documentation, albeit fragmentary, exists on rice systems in West Africa from the 14th century, the discussion in the first section of these systems and their underlying soil and water management principles is based on modern field studies (see Carney, 2001b, for detailed presentation of this archival record). Focus then shifts to the history of rice cultivation in South Carolina, especially during the hundred years from 1670 to 1770, which is crucial since it spans the initial settlement by planters and slaves as well as the expansion of tidal (tidewater) rice cultivation into Georgia. In emphasizing the complex nexus that links culture, technology, and the environment, attention is directed to the indigenous knowledge systems formed in West Africa by ethnic groups speaking Mande and West Atlantic languages. Across the Middle Passage of slavery they brought their expertise with them, and then established rice as a favored dietary staple in the Americas. This knowledge system, moreover, enabled enslaved rice growers to negotiate and alter, to some extent, the terms of their bondage. The concluding section raises several questions about the issue of technology development and transfer, suggests a lingering Eurocentric bias in historical reconstructions of the agricultural development of the Americas, and discusses the scholarly implications of this research.

THE AGRONOMIC AND TECHNOLOGICAL BASIS OF WEST AFRICAN RICE SYSTEMS

Some 4,000 years ago, West Africans domesticated rice along the floodplain and inland delta of the upper and middle Niger River in Mali (Figure 13.2) (Ehret, 2002; Harlan, De Wet, & Stemler, 1976; McIntosh, 1994; Portéres, 1970). The species of rice originally planted in this primary center of domestication, *Oryza glaberrima*, differs from Asian rice, *Oryza sativa*. While both species are currently planted in West Africa, the indigenous African center extends along the coast from Senegal to Côte d'Ivoire and into the Sahelian interior along riverbanks, inland swamps, and lake margins. Within this diverse geographic and climatic setting two secondary centers of *glaberrima* domestication emerged: one, on floodplains of the Gambia River and its tributaries; and another, farther south in the forested Guinea highlands where rainfall reaches 2,000 millimeters/year (Figure 13.2). By the end of the 17th century rice had crossed the Atlantic Basin to the United States, appearing first as a rainfed crop in South Carolina before diffusing along river floodplains and into Georgia from the 1750s.

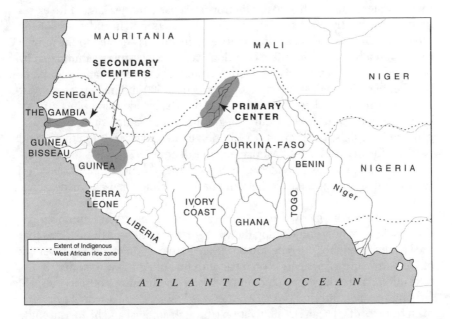

FIGURE 13.2. Centers of origin of African rice (*Oryza glaberimma*). African rice was first domesticated along the floodplain and inland delta of the upper and middle Niger River.

Many similarities characterized rice production on both sides of the Atlantic Basin. In both West Africa and South Carolina the most productive system developed along floodplains. Precipitation in each region follows a marked seasonal pattern, with rains generally occurring during the months from May/June to September/October. Rice cultivation flourished in South Carolina and Georgia under annual precipitation averages of 1,200–1,400 millimeters, a figure that represents the midrange of a more diverse rainfall pattern influencing West African rice cultivation, where precipitation increases dramatically over short distances from north to south (Kovacik & Winberry, 1987). Accordingly, in the Malian primary center and the Gambian secondary center of rice domestication semi-arid (900 millimeters/year) conditions prevail, while southward in Guinea-Bissau and Sierra Leone precipitation exceeds 1,500 millimeters per annum.

The topography of the rice-growing region on both sides of the Atlantic presents a similar visual field. Coastlines are irregularly shaped and formed from alluvial deposits that also create estuarine islands. Tidal regimes on the American coast differ somewhat from those of Africa. The steep descent from the piedmont in South Carolina and Georgia delivers freshwater tidal flows to floodplains just 10 miles from the Atlantic coast. But the less pronounced gradient of rivers in West Africa's rice region means that freshwater tides meet marine water much farther upstream from the coast; on the Gambia River, salinity permanently affects the lower 80 kilometers but intrudes seasonally more than 200 kilometers upstream. Rice cultivation is adapted to the annual retreat and advance of the saline corridor. Even coastal estuaries inundated by ocean tides served as a basis for West African rice experimentation. South of the Gambia River, where precipitation exceeds 1,500 millimeters/year, West Africans learned to plant rice in marine estuaries, by enclosing plots and allowing rainfall to flush out accumulated salts. Water saturation is the key to planting such soils, as it prevents oxidization to an acidic condition that would preclude further cultivation (Moorman & Veldkamp, 1978). An elaborate network of embankments provides a barrier to seawater intrusion, while dikes, canals, and sluice gates enable field drainage of rainwater used for desalination.

Even though higher yielding Asian varieties (*O. sativa*) and pump-irrigation systems are now found along the West African rice coast, the systems that predate the Atlantic slave trade still persevere in the region today. The diversity of West African rice systems is on a par with that of Asia—with some 18 distinct microenvironments planted under diverse combinations of soils, rainfall, farming systems, and land types (Andriesse & Fresco, 1991). In identifying the main features of African rice cultivation (through archival and historical studies as well as contemporary fieldwork), this section examines three principal production systems that bear on the evolution of the Carolina rice plantation economy. The discussion prioritizes water regimes influencing rice cultivation, agronomic techniques, labor demands, and yields in each system.

Rainfall, groundwater, and river tides are the major water regimes influencing West African rice growing, which are termed respectively, "upland," "inland swamp," and "tidal" in this chapter. An African rice system typically involves planting the cereal in numerous microenvironments along a lowland-to-upland landscape gradient with different water regimes, production techniques, and labor utilization (Richards, 1986). This long-standing practice of planting rice sequentially along a landscape continuum confers several advantages, as Dutch geographer Olfert Dapper apprised in the early 17th century (Richards, 1996). The manipulation of multiple water regimes enables farmers to initiate and extend rice growing beyond the confines of a single precipitation cycle. Planting rice in different environments reduces labor bottlenecks since cropping demands (sowing, weeding, and harvesting) unfold in different periods within an agricultural season. By relying on different types of water regimes and cropping environments, farmers enhance subsistence security while minimizing the risk of total crop failure in any given year.

Of the three forms of rice cultivation, only the upland system depends strictly on precipitation; for this reason it is usually the last to be planted (with the onset of rains) in a rice-farming system. "Upland rice" refers to a crop planted at the top of the landscape gradient, with only rainfall for water. West African farmers favor planting upland rice where rainfall is reliable in amount (> 1,000 millimeters/year) and in distribution. The crop is typically grown in an agropastoral system, where wet season rice cultivation rotates with dry season use of the land for cattle pasture. As the livestock feed upon the postharvest stubble, their manure fertilizes the soil for the following season's rice crop. Rain-fed rice generally produces lower yields than rice grown in environments that capture additional water sources. However, the cultivation system is valued throughout the Sahel because it allows a landscape to shift from cereal to livestock production while permitting seasonal access to land by different ethnic groups, thus strengthening food security and protein availability.

Inland swamp cultivation, a second major West African production system, overcomes the constraints posed by seasonal rainfall variability by capturing groundwater from artesian springs, perched water tables, or catchment runoff. "Inland swamps" actually refer to a diverse array of microenvironments, which include valley bottoms, low-lying depressions, and areas where moisture-holding clay soils capture underlying water deposits. These environments often reveal detailed farmer understanding of soils and their moisture retention properties, as well as methods to impound water for supplemental irrigation. For example, by constructing earthen berms around plots, farmers form reservoirs to capture rainfall, which keeps the soil saturated through dry spells in the cropping season. As waterlogging poses a potential cultivation problem, inland swamp plots are often ridged and directly planted with seeds to improve drainage and aeration.

The remaining African production system, tidal rice, occurs on floodplains of rivers and their estuaries. Dependent upon tides to flood and drain fields, tidal cultivation involves many floodplain microenvironments, from those requiring little or no environmental manipulation (freshwater floodplains) to ones demanding considerable landscape modification (cultivation along coastal estuaries, here referred to as "mangrove rice"). Tidal rice cultivation embodies complex hydrological and land management principles that prove especially pertinent for examining the issue of African agency in the transfer of rice cultivation to the Americas.

The cultivation of tidal rice occurs in three distinct floodplain environments: along freshwater rivers, those that are seasonally saline, and coastal estuaries under constant marine water influence. The first two involve similar methods of production—letting river tides irrigate the rice fields—while the third system combines principles of the upland and inland swamp systems for planting the cereal under problematic soil and water conditions. The mangrove rice system received repeated European commentary from the mid-15th century with the arrival of Portuguese caravels to West Africa's rice region (Carney, 2001b). It was depicted and described in detail by a slave captain off the Guinea coast in the 18th century.

Named after the typical vegetation cleared to create the rice landscape, mangrove rice represents the most sophisticated West African production environment. It is found south of the Gambia River, in coastal estuaries flooded by oceanic tides. The key to its success is rainfall regimes where precipitation is abundant and reliable within a cultivation season. Mangroves, whose aerial roots trap alluvium swept over the littoral by marine tides, create these fertile environments. However, their cultivation requires considerable care to prevent the oxidation that would transform them into acid-sulphate soils if no longer submerged in water. By capturing rainfall to leach salt from the field prior to cultivation, then releasing the saline water from the plot with low tides, a rice crop takes advantage of the otherwise extremely fertile soil. After the rice harvest, farmers open up the field's sluices to let marine water enter at high tide until late in the dry season when the water is again drained and the plot prepared for another cultivation cycle.

The technology for rice production in these mangrove estuaries is massive and elaborate—perhaps the reason for early European interest in them. Mangrove rice is also labor-demanding, as the earthen infrastructure of embankments, ridges, and berms must be annually repaired from sloughing and maintained. The embankments, which must be of such a height to prevent the overflow of marine tides onto rice fields, are constructed by hand and often stretch for several kilometers. Dikes to remove the desalinated water and allow dry season entry of ocean tides punctuate the main embankment, while sluices facilitate flooding and drainage of the berms surrounding each plot. Years are involved in shifting a landscape from a mangrove estuary to a rice paddy as the trees are cleared and the area enclosed to reclaim it from the sea.

Over a period of years, the soil is desalinated. Once planting gets underway, controlled flooding with the impoundment of rainfall irrigates the rice plants while drowning unwanted weeds. The reward for such demands on labor is the most productive rice system devised in West Africa.

This overview of the complex soil and water management principles guiding West African rice production illustrates the ingenuity and acumen that from an early date drew the attention of Europeans. As the next section reveals, numerous affinities exist with the rice systems that developed in South Carolina in the early colonial period. The process of technology development in tidewater rice, the antebellum era's quintessential production system, parallels that of the mangrove system in many crucial respects. Scholars of Carolina rice culture have underestimated the active, innovative, and creative role played by the enslaved from West Africa's rice region in the evolution of this cropping system. Transferred by enslaved West Africans to the Americas, imported rice-growing techniques formed part of a larger indigenous knowledge system, grounded along a landscape gradient, or ecological continuum, of microenvironments.

AFRICAN RICE AND AMERICAN CONTINUITIES

By 1860, rice cultivation extended over 100,000 acres along the Eastern Seaboard from North Carolina's Cape Fear River to Florida's St. John's River, and inland for some 35 miles along tidal waterways (Figure 13.3) (Clifton, 1973, 1981). The initial stage of the rice plantation economy dated to the first hundred years of South Carolina's settlement (1670–1770) and, especially, the decades prior to the 1739 Stono slave rebellion. Rice cultivation systems analogous to those in West Africa, as well as identical principles and devices for water control and milling, were already evident in this period. Dramatic increases in slave imports during the 18th century facilitated the evolution of the Carolina rice plantation economy. Technology development unfolded in tandem with the appearance of the task labor system that regulated work on coastal rice plantations. As the crop grew in economic importance, agroenvironments favored for colonial rice production shifted from uplands to inland swamps and, from the 1730s, to the tidal (tidewater) cultivation system that led Carolina rice to global prominence.

This section presents an overview of the historical and geographical circumstances under which rice became a plantation crop in South Carolina. The technical changes marking the evolution of the colonial rice economy illuminate three issues that bear on comparative studies of technology and culture: first, the need to examine the technical components of production as parts of integrated systems of knowledge and not merely as isolated elements; second, the significance of cultural funds of knowledge for technology transfer; and

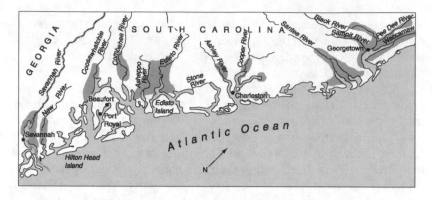

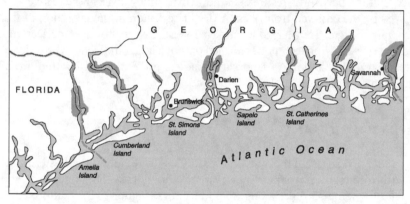

FIGURE 13.3. Tidewater rice cultivation in South Carolina and Georgia.

third, in situations of unequal power relations, the extent to which claims for technological ingenuity can rest on cultural dispossession and appropriation of knowledge.

Slaves accompanied the first settlers to South Carolina in 1670; within 2 years they formed one-fourth of the colony's population; and as early as 1708 they outnumbered whites in the colony (Sirmans, 1986; Wood, 1974). Rice cultivation appears early in the colonial record, with planting already underway in the 1690s. By 1695 South Carolina recorded its first shipment of rice: one and one-quarter barrels to Jamaica (Clifton, 1981). In 1699 exports reached 330 tons, and by the 1720s rice had emerged as the leading trade item (Wood, 1974). Years later, in 1748, South Carolina governor James Glen drew attention to the significance of rice experimentation during the 1690s for development of the colony's economy (Clifton, 1981).

By the 1740s, documents firmly establish the presence of the upland, inland swamp, and tidal floodplain production systems (Carney & Porcher,

1993; Doar, 1936/1970). But rice cultivation in these areas is implied even earlier in the comments of one plantation manager, John Stewart, who claimed in the 1690s to have successfully sown rice in 22 different locations (Wood, 1974). One major point distinguished patterns of land use in West African and South Carolinian rice systems. In West Africa, subsistence security shaped the crop's production, thereby favoring cultivation in numerous microenvironments along a landscape gradient. Rice cultivation in colonial Carolina began as a subsistence crop, planted similarly in diverse environments, but as it became a plantation crop emphasis shifted to specific agroenvironments along the landscape continuum—from rainfed, to inland swamp, to tidal production—to maximize yields and returns on capital and labor.

Upland rice production received initial attention because it complemented the early colony's economic emphasis on stock raising and extraction of forest products. Slave labor buttressed this agropastoral system, which involved clearing forests, producing naval stores (pine pitch, tar, and resin), cattle herding, and subsistence farming (Otto, 1989; Wood, 1974). Export of salted beef, deerskins, and naval stores in turn generated capital to purchase additional slaves. The number of enslaved Africans imported to the colony dramatically increased from 3,000 in 1703 to nearly 12,000 by 1720, which enabled a shift in rice cultivation to the more productive inland swamp system (Gray, 1958; Nairne, 1710/1989; Ver Steeg, 1984; Wood, 1974).

The higher yielding inland swamp system initiated the first attempts at water control in Carolina's rice fields. After clearing swamp forests, slaves developed the network of berms and sluices necessary for converting plots into reservoirs. Like its counterpart in West Africa, the inland swamp system impounded water from rainfall, subterranean springs, high water tables, or creeks to saturate the soil. The objective was to drown unwanted weeds and thereby reduce the labor spent on weeding, as in West Africa. West African principles also guided the cultivation of rice in coastal marshes. Rice was grown in saltwater marshes near the terminus of freshwater streams in soils influenced by the Atlantic Ocean. The conversion of a saline marsh to a rice field depended upon soil desalination, a process not as easily achieved with South Carolina's annual precipitation regime (1,100–1,200 millimeters), which is lower than the average 1,500 millimeters per year that regulate the West African mangrove system. However, by diverting an adjoining freshwater creek or stream, salts could be rinsed from the field (Irving, 1969). The principle of canalizing water for controlled flooding also extended to other settings, such as in places where subterranean springs flowed near the soil surface (Ravenel, 1859). Detailed knowledge of landscape topography and hydrological conditions thus enabled the proliferation of rice growing in diverse inland swamp microenvironments.

During the 18th century rice cultivation in such areas innovated to more elaborate systems of sluices that released reserve water on demand for controlled flooding at critical stages of the cropping cycle (Porcher, 1987; Whitten, 1982). This inland swamp system flourished where the landscape gradient sloped from rain-fed farming to the inner edge of a tidal swamp. Enclosure of a swamp with earthen embankments created a reservoir for storing rainwater, the system's principal source of irrigation. The reservoir fed water, through a sluice gate and canal, by gravity flow to the inland rice field, while a drainage canal and sluice placed at the lower end of the rice field emptied excess water into a nearby stream, creek, or river (Porcher, 1987). Whereas the principle of constructing a reservoir for controlled field flooding is identical to the West African mangrove rice system, the innovative changes that subsequently developed may well provide an instance of what geographer Paul Richards (1996) calls "agrarian creolization." The term refers to the convergence of different knowledge systems and their recombination into new hybridized forms, spearheading the process of innovation.

By the mid-18th century the emphasis on rice had shifted from inland swamps to tidal river floodplains, first in South Carolina, and subsequently in Georgia. The swelling number of slaves directly entering South Carolina from West Africa between the 1730s and the 1770s proved crucial in this spatial relocation of the rice economy. Some 35,000 slaves were imported into the colony during the first half of the century and over 58,000 between 1750 and 1775, making South Carolina the largest importer of enslaved Africans on the North American mainland between 1706 and 1775. The share of slaves brought directly from the West African rice coast grew during these crucial decades of tidewater rice development from 12% (1730s) to 54% (1749–1765), and then to 64% (1769–1774) (Richardson, 1991). This pattern is illuminated in a typical handbill from the colonial period, which announces the sale in Charlestown (Charleston) on July 24, 1769 of enslaved men, women, and children from Sierra Leone (Figure 13.4).

Tidewater cultivation occurred on floodplains along tidal rivers where, similar to its mangrove rice counterpart in West Africa, the diurnal variation in sea level facilitated field flooding and drainage. Preparation of a tidal floodplain for rice cultivation followed principles already outlined for the mangrove rice system. The rice field was embanked at sufficient height to prevent tidal spillover, while the earth removed in the process created adjacent canals. Sluices built into the embankment and field sections operated as valves for flooding and drainage. The next step involved dividing the area into plots (in South Carolina these were termed "quarter sections," of some 10–30 acres), with river water delivered through secondary ditches. This elaborate system of water control enabled the adjustment of land units to labor demands and allowed slaves to directly sow rice along the floodplain (Clifton, 1981). Then the rice was planted directly in the floodplain, as it is in African floodplain cultivation.

Tidewater cultivation required considerable landscape modification and

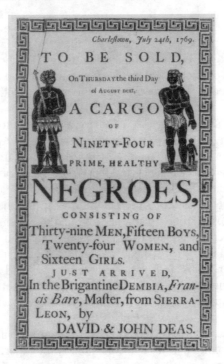

FIGURE 13.4. Handbill announcing a sale of slaves from Sierra Leone in Charleston, South Carolina, July 24, 1769. Reprinted courtesy of the Library of Congress.

ever-greater numbers of enslaved laborers than rain-fed and inland swamp cultivation. Leland Ferguson, a historical archaeologist, vividly captures the staggering human effort involved in transforming Carolina's tidal swamps to rice fields:

> These fields are surrounded by more than a mile of earthen dikes or "banks" as they were called. Built by slaves, these banks . . . were taller than a person and up to 15 feet wide. By the turn of the eighteenth century, rice banks on the 12½ mile stretch of the East Branch of Cooper River measured more than 55 miles long and contained more than 6.4 million cubic feet of earth." By 1850, aided only by hand-held tools, slaves in the Carolina rice zone had built earthworks "nearly three times the volume of Cheops, the world's largest pyramid." (Ferguson, 1992, xxiv–xxv)

While such landscape change placed considerable demands on slave labor for construction and maintenance, it reduced the need for manual weeding, one of the most labor-intensive tasks in rice production. The systematic lifting and lowering of water was achieved by sluices, known as "trunks," located in the embankment and secondary dikes; by the late colonial period these devices had evolved into hanging floodgates (Porcher, 1985). With full control of an adjacent tidal river, the rice field could be flooded on demand for irrigation and weeding and to renew the soil annually with alluvial deposits. Because of this increasing reliance on water control technology, in tidal cultivation one slave could manage 5 acres of rice, as opposed to just 2 acres in the inland rice system (Clifton, 1981; Gray, 1958).

The history of the term "trunk" for sluice gate in Carolina also suggests evidence for technology transfer from the West African rice coast. While the hanging gate technology likely provides another example of agrarian creolization, its name refers to an earlier device. Even when the hanging gate replaced earlier forms, Carolina planters continued to call sluice gates "trunks." In the 1930s planter descendant David Doar stumbled upon the term's origin: the earliest sluice gates were formed from hollowed-out cypress trunks (Doar, 1936/1970). The original Carolina sluice system looked and functioned exactly like its African counterpart. Reference to them as "trunks" throughout the antebellum period suggests that the technological expertise of Africans again proved significant in Carolina rice history.

Tidewater rice cultivation led South Carolina to global economic prominence in the 18th century. It was made possible by the expertise of enslaved Africans in cultivating as well as processing the crop. Their experience with planting a whole range of interconnected environments along a landscape gradient likely permitted the sequence of adaptations that marked the growth of the South Carolina rice industry.

African contributions to Carolina rice history were not limited to cultivation practices. They also extended to the method by which the grain was processed for consumption. Until suitable mechanical mills were developed around the time of the American Revolution, the entire export crop was milled by hand, in the traditional method long used by African women, with a mortar and pestle. Even the fanner baskets used to winnow rice on Carolina and Georgia plantations display a likely African origin, as anthropologist Dale Rosengarten suggests. She links the coiled-basket-weaving tradition in fanner baskets to the Senegambian rice region, as it did not exist among Native Americans of the Southeast region (Rosengarten, 1999).

Enslaved Africans thus contributed significantly more than physical labor to colonial rice production. They provided the critical expertise in establishing rice cultivation in South Carolina, even though accounts by planters and their descendants have long claimed for their forebears the ingenuity in developing the system. Planter accounts, however, fail to explain how they learned to transform wetland landscapes by careful observance of tidal dynamics, soils, microenvironments, and water regimes. Such systems of planting cereals in standing water did not exist in England at the time, yet surfaced within two decades of Carolina's settlement. Wetland rice represents a far more complex farming system than the rainfall cultivation practiced by the English. Enslaved Africans from the African rice region were thus the only settlers present in the Carolina colony who possessed this knowledge system.

The delayed recognition of this significant contribution to the history of the Americas stems, in part, from the way that rice has been examined by scholars. By emphasizing rice as a cereal, as a grain consumed and traded internationally, studies have failed to place it within its proper agroecological context for studies of diffusion and technology transfer. An examination of rice as a landscape of microenvironments brings into focus the underlying soil and water management regimes that inform its cultivation as well as its cultural origins. Recovery of the African contribution to American rice history also involves considering consumption as well as production, processing as well as cultivation, and the types of labor and knowledge that mediated the spatial movement of rice from field to kitchen. This requires sensitivity to ecological as well as gendered forms of knowledge in technology transfer.

While this overview of rice beginnings in South Carolina argues that planters reaped the benefits of a rice-farming system perfected by West Africans over millennia, an important question remains. Why would enslaved West Africans transfer a sophisticated rice system to plantation owners when the result spelled endless and often lethal toil in malarial swamps?

The answer is perhaps revealed in the appearance by 1712 of the task labor system that characterized coastal rice plantations (Morgan, 1972, p. 565). It was distinguished from the more pervasive "gang" form of work typifying plantation slavery. In the gang system, "the laborer was compelled to work the entire day," while "under the task system the slave was assigned a certain amount of work for the day, and after completing the task he could use his time as he pleased" (Gray, 1958, pp. 550–551). Without underestimating the real toil involved in the two systems, the task system did set normative limits to daily work demands. Such seemingly minor differences between the two systems could deliver tangible improvements in slave nutrition and health, as Johan Bolzius implied in 1751 with his observation: "If the Negroes are Skilful and industrious, they plant something for themselves after the day's work" (quoted in Morgan, 1972, p. 565).

The task labor system appeared at the crucial juncture of the evolution of rice as a plantation crop in the Carolina colony and the shift to the more productive, but labor-demanding, inland swamp system. A similar system of limiting demands placed on enslaved labor was already in existence along Africa's Rice Coast (Carney, 2001b). The appearance of the task labor system in Carolina's fledgling rice economy may well represent the outcome of negotiation and struggle between master and slave over knowledge of rice culture and the labor process to implement it. In providing crucial technological acumen, slaves perhaps discovered a mechanism to negotiate improved conditions of bondage. But by the 19th century such gains had eroded, as the frontier for slave escape closed and slavery appeared to be a permanent feature of Southern agriculture. The task labor system became little different than the gang form of slavery (Carney, 2001b).

CONCLUSION

"What skill they displayed and engineering ability they showed when they laid out these thousands of fields and tens of thousands of banks and ditches in order to suit their purpose and attain their ends! As one views this vast hydraulic work, he is amazed to learn that all of this was accomplished in face of seemingly insuperable difficulties by every-day planters who had as tools only the axe, the spade, and the hoe, in the hands of intractable negro men and women, but lately brought from the jungles of Africa" (Doar, 1936/1970, p. 8). In 1936, when David Doar, descendant of Carolina planters, echoed the prevailing view that slaves contributed little besides labor to the evolution of the South Carolina rice economy, no historical research suggested otherwise. While recent research challenges such unquestioned assumptions, a bias nonetheless endures against considering West Africans as the originators of rice culture in the Americas.

Even authoritative texts on rice cultivation, such as that of D. H. Grist (1968), express such a bias. In reference to a type of paddy rice cultivation found in British Guiana (now Guyana) and the neighboring former Dutch colony of Surinam, Grist describes the "empoldering" technique as "a method of restricting floods and thus securing adjacent areas from submergence" (Grist, 1968, p. 45). While this technique is strikingly similar to that employed in mangrove rice production along the West African rice coast, he attributes the system to 18th-century Dutch colonizers (Grist, 1968). In that era, however, Surinam possessed one of the highest ratios of Africans to Europeans of any New World plantation society (65:1 in Surinam compared to Jamaica's 10:1) (Price & Price, 1992).

More recent work in Brazilian rice history repeats this perspective. Even though rice was not grown in Portugal during the colonization of the Americas, Pereira (2002) attributes its 16th-century introduction in Brazil's eastern Amazon and Northeast to migrants from the Azores and Portugal. There is no discussion of how the African mortar and pestle (a device not used in Portugal) came to be the sole technology for milling rice in Brazil until the mid-18th century, when it was replaced by water mills that successfully removed the hulls without grain breakage. Nor is there any acknowledgment of the possibility that rice culture became established in Brazil, as in South Carolina, because African rice routinely provisioned slave ships, providing the enslaved an opportunity to grow their food staple for subsistence (Carney, 2001a).

Evidence from the American and African Atlantic thus suggests that slaves from West Africa's indigenous rice area established rice culture. A crop initially planted for subsistence became, in 18th-century South Carolina, the first cereal globally traded as a plantation export crop. This complex indigenous knowledge system guided the transformation of Carolina's swamps while serving as a source of technological innovation in rice culture, captured in the notion of agrarian creolization. These achievements in tidal rice diffused southward in the 18th century along rivers in Georgia and Florida, and, from midcentury, overseas to similar environments in Brazil's Northeast and eastern Amazon regions. While rice systems of the Americas would eventually bear the imprimatur of both African and European influences, its appearance initially is linked to a knowledge system developed in West Africa and carried across the Middle Passage by slavery's victims.

Thus, as Europeans and Africans faced each other in a new territory under dramatically altered and unequal power relations, the enslaved established a subsistence crop long valued in West Africa. With the abolition of slavery, rice history led to cultural dispossession and appropriation by descendants of slave owners who credited the beginnings of rice farming to European ingenuity and presumed mastery of technology. A careful reading of the archival and historical record—one attuned to agroenvironments, power relations, ecological principles, and social history—reveals a dramatically different narrative.

The ending of this story is not yet settled: this cross-cultural, social–environmental history of transatlantic rice culture also cuts across current political and cultural debates. In the U.S. South, historical preservation of antebellum landscapes is charged with the controversy of memory politics. While many descendants of white planters view preservation of plantation landscapes as a source of pride, many African Americans perceive these as sites of toil, terror, and shame. This discussion of rice technology transfer demonstrates that despite the brutality of bondage, African slaves were active and ingenious shapers of antebellum agroenvironments rather than mere physical laborers. Historical preservation and ecological restoration efforts in the South should acknowledge the rich hybrid nature of these cultural landscapes in a way similar to that currently underway in the reconsideration of southern family histories by black and white descendants of slave owners.

This study also has implications for the hotly debated issue of intellectual property rights over agricultural seeds. Today, cultivars and their germ plasm are increasingly engineered, patented, and privatized. This market logic reduces seeds to their mere biology. As this investigation of rice cultivation on the Atlantic Rim implies, however, seeds cannot be so easily separated from their political and social context. The introduction of African rice in the New World was not simply the movement of seeds from one environment to another, but rather the transfer and transformation of a *rice culture*, with attendant continuities and changes in technology, labor organization, social structures, and cultural meanings.

ACKNOWLEDGMENTS

I would like to thank Eric Carter, as well as the book's editors, for their invaluable comments on this chapter.

REFERENCES

Andriesse, W., & Fresco, L. O. (1991). A characterization of rice-growing environments in West Africa. *Agriculture, Ecosystems and Environment, 33,* 377–395.

Carney, J. (2001a). African rice in the Columbian Exchange. *Journal of African History, 42*(3), 377–396.

Carney, J. (2001b). *Black rice: The African origins of rice cultivation in the Americas.* Cambridge, MA: Harvard University Press.

Carney, J., & Porcher, R. (1993). Geographies of the Past: Rice, Slaves and Technological Transfer in South Carolina. *Southeastern Geographer, 33*(2), 127–147.

Clifton, J. (1973). Golden grains of white: Rice planting on the lower Cape Fear. *North Carolina Historical Review, 50,* 365–393.

Clifton, J. (1981). The rice industry in colonial America. *Agricultural History, 55,* 266–283.

Doar, D. (1936/1970). *Rice and rice planting in the South Carolina low country.* Charleston, SC: Charleston Museum.

Egan, D., & Howell, E. A. (Eds.). (2001). *The historical ecology handbook.* Washington, DC: Island Press.

Ehret, C. (2002). *The civilizations of Africa: A history to 1800.* Charlottesville: University Press of Virginia.

Fairhead, J., & Leach, M. (1996). *Misreading the African landscape.* Cambridge, UK: Cambridge University Press.

Ferguson, L. (1992). *Uncommon ground: Archaeology and early African America, 1650–1800.* Washington, DC: Smithsonian.

Flanders, R. B. (1933). *Plantation slavery in Georgia.* Chapel Hill: University of North Carolina Press.

Gray, L. (1958). *History of agriculture in the southern U.S. to 1860* (2 vols.). Gloucester, MA: Peter Smith.

Grist, D. H. (1968). *Rice* (4th ed.). London: Longmans, Green.

Harlan, J., De Wet, J., & Stemler, A. (1976). *Origins of African plant domestication.* Chicago: Aldine.

Heyward, D. (1937). *Seed from Madagascar.* Chapel Hill: University of North Carolina Press.

Irving, J. B. (1969). *A day on the Cooper River.* Charleston, SC: R. K. Bryant.

Kovacik, C., & Winberry, J. (1987). *South Carolina: The making of a landscape.* Boulder, CO: Westview Press.

Linares, O. F. (1992). *Power, prayer and production.* New York: Columbia University Press.

Littlefield, D.C. (1981). *Rice and slaves.* Baton Rouge: Louisiana State University Press.

McIntosh, S. K. (1994). Paleobotanical and human osteological remains. In S. K. McIntosh (Ed.), *Excavations at Jenne-jeno, Hambarketolo and Kaniana in the Inland Niger Delta (Mali): The 1981 season* (pp. 348–353). Berkeley and Los Angeles: University of California Press.

Moorman, F. R., & Veldkamp, W. J. (1978). Land and rice in Africa: Constraints and potentials. In I. Buddenhagen & J. Persely (Eds.), *Rice in Africa* (pp. 29–43). London: IRRI.

Morgan, P. (1972). Work and culture: The task system and the world of low country blacks, 1700 to 1880. *William and Mary Quarterly,* 3rd ser., *39,* 563–599.

Nairne, T. (1710/1989). A letter from South Carolina. In Jack Greene (Ed.), *Selling a new world: Two colonial South Carolina promotional pamphlets* (pp. 33–73). Columbia: University of South Carolina Press.

Otto, J. S. (1989). *The southern frontiers, 1607–1860.* New York: Greenwood Press.

Pereira, J. A. (2002). *Cultura do Arroz no Brasil.* Teresina, Piauí: EMBRAPA.

Porcher, R. (1987). Rice culture in South Carolina: A brief history, the role of the Huguenots, and preservation of its legacy. *Transactions of the Huguenot Society of South Carolina, 92,* 11–22.

Portéres, R. (1970). Primary cradles of agriculture in the African continent. In J. D. Fage & R. A. Oliver (Eds.), *Papers in African prehistory* (pp. 43–58). Cambridge, UK: Cambridge University Press.

Price, R., & Price, S. (1992). *Stedman's Surinam: Life in an eighteenth-century slave society.* Baltimore: Johns Hopkins University Press.

Ravenel, E. (1859, Feb. 1). The limestone springs of St. John's, Berkeley. *In Proceedings of the Elliott Society of Science and Art of Charleston, South Carolina,* 28–32.

Richards, P. (1986). *Coping with hunger.* London: Allen & Unwin.

Richards, P. (1996). Culture and community values in the selection and maintenance of African rice. In S. Brush & D. Stabinsky (Eds.), *Valuing local knowledge: Indigenous people and intellectual property rights* (pp. 209–229). Washington, DC: Island Press.

Richardson, D. (1991). The British slave trade to colonial South Carolina. *Slavery and Abolition, 12,* 125–172.

Rosengarten, D. (1997). *Social origins of the African-American lowcountry basket.* Unpublished doctoral dissertation, Harvard University, Cambridge, MA.

Sirmans, E. (1986). *Colonial South Carolina: A political history, 1662–1763.* Chapel Hill: University of North Carolina Press.

Smith, J. F. (1985). *Slavery and rice culture in low country Georgia 1750–1860.* Knoxville: University of Tennessee Press.

Ver Steeg, C. (1984). *Origins of a southern mosaic.* Athens: University of Georgia.

Whitten, D. (1982). American Rice Cultivation, 1680–1980: a tercentenary critique. *Southern Studies, 21,* 5–26.

Wood, P. (1974). *Black majority.* New York: Knopf.

CHAPTER 14

Future Directions in Political Ecology

Nature–Society Fusions and Scales of Interaction

Karl S. Zimmerer
Thomas J. Bassett

This book covers a wide range of issues of an entwined social–environmental nature that range from water and biodiversity resources to ideas of nature and management policies and scientific concepts for conservation. The political ecology approach that guides these chapters is evolving rapidly through the contribution of innovative works in many areas. In concluding, our discussion is focused on future directions that are represented here and that are being forged in the debates, discussions, and research that surround and inform these efforts. Concluding assessments are divided, first, among the five main topical areas (protected areas, urban and industrial environments, ecological analysis and theory in resource management and conservation, geospatial technologies and knowledge, and North–South environmental histories). The second concluding section is concerned with future directions that correspond to the book's central themes of fused nature–society analysis and the political ecology of scale.

A pair of general, deeper level questions has underlain the chapters in this book which are also evident in the ensuing discussion of future directions. First, what role do geographic differences play in the future of political ecology? The initial emphasis of the political ecological approach was on rural societies and resources, often soils and agriculture, which were located primarily in developing countries of the global South. Increasingly, the geographic situations under consideration are commonly broadened to include the global North and urban and industrial settings. Diverse environmental and resource issues now range from human health and disease to the roles of nonprofit or nongovernmental organizations (NGOs). This shift marks an important and welcome trend in geographic political ecology. At first glance, the new attention to a fully global range of topics appears to closely resemble the viewpoint of "ecological modernization" that highlights the expanded role of environmental management and technomanagerial solutions such as government regulations, legal frameworks, and protected areas (Buttel, 2000).

Yet, as was described in Chapter 1, much uncertainty concerns the geographical dynamics that are associated with globalization of environmental management, particularly in the global South. In various ways all the chapters offer evidence, varied yet abundant, of the dissemination and heightened profile of global environmental management and associated laws and regulatory institutions. These chapters also show, however, that these changes in the global South do not at all mimic the style of environmental reform in advanced industrial societies. Not surprisingly, the priorities of many global environmental managers, such as biodiversity and ecosystem management, simply do not match closely with what matters most to a large majority of the rural and urban inhabitants of newly industrialized and developing countries. Even more important, the globalization of environmental issues and the spread of legal and institutional frameworks are taking shape quite differently "on the ground" due to the far different situations of the global South. In short, the significance of geographic places and the differences among them are being made more important, rather than less so, amid the globalization of environmental change.

Second, what is the role of geography as a discipline in the future of political ecology? This question is central to the material presented in this book, and it is relevant in addressing the issues that are presented as future directions below. On the one hand, *interdisciplinarity* is clearly a cornerstone of contemporary environmental studies. Transdisciplinarity is also gaining ground (Fry, 2001). Changes in political–ecological conditions are plainly multifaceted in nature. Incorporating the multisided aspects of biogeophysical and social dimensions makes urgent the call for integration. The preeminence of environmental management as a social goal further adds to this impetus. For example, environmental management in government and scientific circles is founded on the advice and analysis of teams of scientific experts. Indeed, the impulse of environmental reform is leading to the transformation of institutions for the production and dissemination of knowledge, seen, for example, in the interdisciplinary programs of the environmental sciences and environmental studies in academic institutions (Pezzoli, 1997).

On the other hand, it is also the case that interdisciplinary scope must be accompanied by disciplinary depth and rigorous transdisciplinary synthesis. As argued below, the discipline of geography must as a consequence develop a more probing discussion of its place amid the growing and diversifying field of political ecology. Our analysis of geography's contribution as a discipline is also relevant to identifying its role amid the more general raising tide of inter-disciplinarity and transdisciplinarity. The magnitude of this challenge to geography is suggested in the most recent presidential address of the Association of American Geographers (Golledge, 2002). In exploring this area, our book adopts the view that this consideration of disciplinarity in political ecology with regard to environmental issues is akin to disciplinary discussions on such fields as development studies and gender and women's studies (Castle, 2002; Friedman, 2001; Jackson, 2002).

The transdisciplinarity of this volume's approach to environmental issues can be summarized in the observation that, as demonstrated in all the chapters, the authors are inveterate weavers of analysis that repeatedly bridge the social and biogeophysical sciences. Inveterate weaving means that the integration across this divide occurs numerous times in a close-knit fashion, rather than being segregated in the research design or write-up (e.g., by the sections of the analysis, as is typical of many interdisciplinary reports). Tight inter-weaving also reflects that transdisciplinary research is dependent not only on disciplinary rigor but also on "scientific connoisseurship" in one or more other disciplines (Rayner, 2003). In other words, the use of these other perspectives must be well-informed rather than perfunctory. We believe that the style of interweaving undertaken in this volume is relevant to other types of political ecology. By focusing on one type of interdisciplinarity, our intent is to engage not just this specialty but also the many transdisciplinary practitioners within political ecology whose connections may trace more to sociology, anthropology, planning, environmental history, or environmental resource management studies.

Close interweaving of disciplinary perspectives is also evident in the consideration of social and political causation that is related to management and change of natural environments. The chapters' careful attention to this causation is in contrast to the list-making of factors, a method that only nods to connections or vaguely imputes the role of a general condition such as poverty. An example of the later would be unspecified reference to victimization or oppression (see Vayda & Walters, 1999), which is one symptom of the "academic hitchhiker's" view of political ecology (Blaikie, 1998). Finally, the weaving of socioenvironmental analysis in our volume points to the prominence of scale. The chapters indicate the varied range of the scales of socioenvironmental interaction. This range is not predetermined but rather emerges from the scaling processes (e.g., urban neighborhoods, single or multiple rural communities, watersheds, protected areas and subunits within them). Our volume demonstrates that many of these scales are recognized among diverse disciplines. Such shared recognition can offer a basis for our attempts at integration and broaden the current emphasis on the landscape scale (Fry, 2001).

ADVANCES IN FUTURE DIRECTIONS: CORE TOPICS

Protected Areas

A recurring theme of political–ecological research on protected areas is the tension that exists between local and global agendas for "sustainable development." Over the past couple of decades, two global summits on environment and development[1] have shown that a diversity of actors and institutions have sometimes conflicting, sometimes complementary, interests in how land and water resources are managed (Goldman, 2002; Laurie, Andolina, & Radcliffe, 2002). While the World Bank encourages the privatization of everything in its development policies (Watts, 1994), grassroots organizations mobilize to defend what they consider to be ancestral if not God-given rights to land and water.

Similar tensions play out in the arena of biodiversity conservation. International environmental NGOs like the World Wide Fund for Nature (WWF) and the International Union for the Conservation of Nature (IUCN) team up with multilateral institutions like UNESCO and the World Bank to designate particularly biodiverse zones as protected areas. These conservation territories are imposed on local communities whose livelihoods depend on access to the resources now declared off-limits. To counter the widespread practice of coercive conservation (Moore, 1993; Neumann, 1998, 2001; Peluso, 1996), scholars and grassroots activists are seeking alternative conservation agendas in which the rights of peoples inhabiting biodiverse areas are respected. These calls for environmental justice need to be supported by scholarly work that focuses on what Zerner (1998, p. 16) calls "regimes of nature management" in which "certain species, landscapes, and environmental outcomes are privileged while others are peripheralized or disenfranchised."

In this volume, Young (Chapter 2) and Sundberg (Chapter 3) map out a number of research directions centered on nature management regimes. The first focuses on the designation of patrimony or the natural heritage that is the object of conservation or protection. Young and Sundberg examine the criteria of selection and environmental representations produced by NGOs, multilateral banks, and local peoples. Their work shows how NGOs, tourist companies, and state agencies privilege specific ecosystems, species, and landscapes as being more important than others (Cormier-Salem, Juhé-Beaulaton, Boutrais, & Roussel, 2002). They contrast global and local values and criteria guiding natural resource management, the divergent and complementary interests at work, and the uneasy alliances among diverse actors that emerge around the designation of patrimony.

A second and related research direction concerns the identity politics of natural resource conservation and preservation. In the Petén peninsula of Guatemala, international conservation NGOs exclude certain groups (ranchers, loggers, oil companies) from their programs while appealing to others (*Peteneros*) who are viewed as authentic forest dwellers who are most likely to carry out NGO agendas. Sundberg describes how struggles between competing forest user groups over rights to resources is articulated in terms of cultural identities that are refashioned in the context of conservation discourses and programs.

Third, the chapters offer insights into how markets (their configuration, scale, dynamics) and patrimony intersect in potentially innovative ways. In the context of a depleted fishery, Young examines the viability of ecotourism as a method of biodiversity conservation. In other areas of Mexico, certified organic coffee growing is emerging as a novel market outlet for smallholder coffee growers that combines niche markets and environmental conservation (Mutersbaugh, 2002b). Can other regional products be promoted that will boost rural incomes and at the same time be conducive to resource conservation? Political ecologists are at the forefront in examining regimes of nature management that link rural livelihoods with multiscale political–economic and ecological processes.

Urban and Industrial Environments

The chapters by Pelling (Chapter 4) and Swyngedouw (Chapter 5) point to a number of future directions in the political ecology of urban and industrial environments. In particular, their contributions show a series of promising intersections with three areas of rapidly accelerating interest: (1) environmental justice and politics; (2) the evolving social–environmental role of risks, the adjustments that are made in response to them, and the mix of technoscientific and citizen knowledges that are involved; and (3) geographic frameworks of linked urban–rural environments that seek to model and build on new images of this dynamic continuum. In the following discussion, these directions are evaluated with respect to how the advances in political ecology can engage productively with larger debates and issues.

Environmental justice, both as an interdisciplinary intellectual endeavor and an increasingly effective umbrella for social movements and activism, holds much promise for a broader engagement with political ecology. To date, the general approach of environmental justice is associated primarily with advanced industrial and urban settings where low-income and minority communities have been disproportionately subjected to toxic contaminants in areas of hazardous waste disposal and other heavily polluted sites (Dichiro, 1998; Hurley, 1995). In the United States, the location of a toxic waste landfill in a low-income, African American community in North Carolina provided a major impetus for the environmental justice movement. Thus far the perspective of environmental justice is focused primarily on activist political organizing and descriptive historical narratives. As a result, it seems poised for fruitful engagement with a more conceptual analysis of the geographically uneven social–environmental dynamics of development. Here the works of Pelling and Swyngedouw offer vivid examples using case studies of water resource control and risk. Their chapters are part of political ecology's growing integration of theoretical ideas of the uneven social–environmental dynamics of development and the perspective of environmental justice (Martinez-Alier, 2000; Rocheleau, Thomas-Slayter, & Wangari, 1996; Schroeder, 2000; Stonich, 1995). Expanded analysis of the evolving geographic unevenness of political–ecological change, which seems etched with particular prominence in urban industrial settings, is thus already adding to a fuller perspective on environmental justice (and vice versa).

The idea of environmental risk and social responses continues to evolve as a touchstone of thinking on nature–society relations in the works of Pelling and Swyngedouw and in other recent contributions (e.g., Kirkby, O'Keefe, & Howorth, 2001). It seems especially salient in urban-industrial settings where environmental and technological risk is being viewed as fundamental to the organization of present-day and future societies (Beck, 1992, 1999). The rise of a so-called risk society is illustrated mostly with examples of the urban industrial milieu. This area of political ecology is primed for potentially fruitful exchanges with the proliferating literature on risk society (Bryant & Bailey, 1997; Watts, 2000). The works of Pelling and Swyngedouw suggest important contributions to this dialogue. Their chapters demonstrate that environmental risk is being actively produced and managed, and that it is dynamically uneven in modern and late-modern stages of city and rural development.

Urban–rural linkages of resource management are amply illustrated in the chapters of Pelling and Swyngedouw. Water demands of cities and industry exert a mounting pressure on rural water users, and vice versa. Likewise the biogeophysical aspects of flooding risk within cities are shaped strongly through an urban–rural continuum of river channel and habitat modifications. Pelling and Swyngedouw point to such expanding urban–rural linkages and, by doing so, suggest how they can be viewed via geographic frameworks of urban–rural linkages. Their framings of the urban–rural continuum echo the geographic linking that is evidenced in examinations of the effects of global or transnational processes on agriculture, ranging from growing under agroindustry contracts to the transnational certification of organic agriculture (Jarosz & Qazi, 2000; Mutersbaugh, 2002a, 2002b; Watts, 1998; Watts & Goodman, 1997).

By contrast, the issues of urban and rural locales are still seen as largely apart in the realm of environmental ideas and conservation. They are commonly separated into the "Green Agenda" of rural wilderness- or resource-based conservation and the "Brown Agenda" of urban environmentalism that targets air and water pollution, toxic contaminants, and sprawl. This persistent conservationist tendency of cleaving countryside and city is a powerful legacy of romanticism. Yet Pelling and Swyngedouw show that urban–rural linkages can often serve as central components of the frameworks for the analysis of resource use and management. Their works suggest that analytical framing of this environmental continuum is a promising future direction for the formulation of geographic models and ideas that are expressed in cultural and political images of "geographic imaginaries" (McCarthy, 2002; Zimmerer, 2000).

Ecological Analysis and Theory in Resource Management and Conservation

Future directions of ecological analysis and the incorporation of ecological science are highlighted in the chapters by Bassett and Koli (Chapter 6), Zimmerer (Chapter 7), and Turner (Chapter 8). Their works underscore various dimensions in this area that may serve as promising avenues for geographical political ecology. Prominent among these aspects are (1) the examination of human–ecological concepts qua concepts; (2) contributing to the expanded undertakings of the human–environmental and global-change scientific communities; and (3) addressing the practice and politics of ecological science in development institutions.

One noteworthy commonality of the works of Bassett and Koli, Zimmerer, and Turner is their interest in the scientific meanings of human-ecological concepts as analytical constructs. Their chapters parlay this interest into the testing, evaluation, and refinement or reformulation of ecological scientific concepts that are common and often key to the analysis of human–environment interaction. Investigating the apparent increase of woodlands in Côte d'Ivoire, for example, Bassett and Koli are focused on the ecological scientific concept of plant succession. They identify a common successional pathway that leads from grassland toward open forest vegetation and evaluate the roles of human–environment activities that entail recent changes in burning, farming, and grazing. Similar in this sense, Zimmerer's chapter is focused on the ecological concept of the zonation of agricultural land use. Turner's work is centered on that of the carrying capacity of rangelands. As demonstrated in these chapters, a variety of mainstay human–ecological concepts may continue to yield substantial revisions and refinement or potentially reformulation. Sustained and systematic testing and evaluation, framed through the perspective of political ecology, holds a promising future in this endeavor.

A second future direction involves the growing size and political role of national and international scientific communities. One such community, which at first glance suggests a resemblance to political ecology, consists of scientific experts that operate under the umbrella of global human–environmental and global-change concerns. Among these specialists is the community that has identified itself as concerned with the global-change analysis of land use and land cover, such as occurs, for example, in the Land-Use and Land-Cover Change (LUCC) group (Lambin et al., 2001; Turner et al., 2001). Allied to other global environmental modeling, of climate and carbon in particular, LUCC is proving most capable, thus far, in the analysis through the use of remote sensing, GIS, and census reports of the "what" aspect of global changes in agriculture and vegetative cover. Gaining adequate understandings of the "how" and "why" aspects of these dynamics of change are important areas where opportunities may abound for political ecology (Bassett & Crummey, 2003; Homewood et al., 2001).

The potential contribution of geographical political ecology to the scientific community of LUCC is illustrated in the widespread occurrence of such political–ecological processes as the influential relations of ethnicity–social power–agricultural and livestock markets (chapters by Bassett and Koli and Turner), and also the household resource diversification–land use–community institutions (Zimmerer chapter). Political–ecological analyses of such relations of "how" and "why" issues are likely to offer a more rigorous explanation than the narrowly demographic or market-signal assumptions of so-called drivers of change. Although the latter factors are often abstracted from the actual explanation of environmental change itself, they tend to fulfill the requirements of many existing models for readily quantifiable data on "human dimensions" that are available for large areas. It may be that in the near future the political ecology analysis will exist at best in a productive tension with the global-environmental modeling and "drivers of change" approaches (Klooster & Masera, 2000; Turner, 1999). Qualitative and quantitative modelling of household- and community-based resource management is already a source of productive overlap in these approaches (Bassett, 2002; Coomes, 1996; Coomes & Barham, 1997).

Ecological science continues to expand worldwide within the context of development institutions where it is a source of information and a claim to power and influence. This topical area is primed for future contributions from geographical political ecology. Possible directions may be seen as extending from the examples that are offered in the chapters of Bassett and Koli, Zimmerer, and Turner. Indeed, the lens of development institutions shows that each chapter is concerned with the ideas, practice, and implementation of ecological science in such agencies as the World Bank (Bassett and Koli), agrobiodiversity conservation plans, and programs such as those undertaken in international UNESCO Biosphere Reserves and among international and local NGOs (Zimmerer); and range management and conservation organizations within national governments and international aid groups (Turner). These chapters adopt a view of institutional ecological science in these development agencies as both discursive and, at the same time, scientifically materialist in makeup. It is discursive in its production and use for institutional purposes, while it is materialist in it philosophical claims to scientific truths that are epistemologically committed to nearest approximations of reality. The dual makeup of science as simultaneously discursive-materialist and its expanding use among development institutions are features that fit well with geographical political ecology and the theme of entwined nature–society interaction. It is a promising focus for the future. Political ecology can advance the critical epistemological awareness that is needed in order to move beyond scientific myths and error that tend to show a remarkable persistence due to the interaction of vested power, interests, and mind-sets.

Geospatial Technologies and Knowledges

Robbins (Chapter 9) and McCusker and Weiner (Chapter 10) suggest that one productive way to integrate geospatial technologies and political ecological research is through participatory GIS. In the global North, participatory GIS refers to interactive mapping on the Internet. In the global South, it refers to the involvement of local groups in the multiple stages of image interpretation and map making. The difference in definitions speaks volumes about the expert nature, accessibility, and power relationships embedded in GIS technologies. The more general point made by Robbins and by McCusker and Weiner is that the interpretation of landscapes and participatory mapping are inherently political processes. How might these technologies be employed so that they empower rather than further subjugate subordinate groups? One challenge lies in focusing more closely on the different stages in which a GIS is constructed,

Two stages especially stand out where vulnerable stakeholders (e.g., landless groups, women, poor households) might meaningfully intervene: the problem definition stage and the information stage. McCusker and Weiner systematically solicit the views of vulnerable groups in defining the problems to which GIS might be applied. If the pressing issue for the majority of the rural population in South Africa is land redistribution, then how can GIS be used to further this goal? Robbins points to the critical importance of the information stage. Deciding what constitutes forests is clearly central to subsequent land cover mapping and whose interests are served in the process. In the absence of GIS stations in remote rural areas, the involvement of socially marginalized groups in these critical stages in the GIS construction is essential if these technologies are to improve their livelihoods. More research needs to be conducted on how the diffusion of geospatial technologies actually empowers and improves the lives of the poor.

The scale of analysis is a recurring theme in this section. McCusker and Weiner argue that small-scale images (e.g., 1 kilometer × 1 kilometer pixels) naturalize the landscape by removing traces of human–environmental history. On the other hand, very-large-scale images (e.g., 4 meter × 4 meter pixels) produce too much data. The scale of analysis can also be constrained by data structure. For example, a GIS is often structured by a census enumeration area rather then by the political, economic, or ecological dynamics shaping the landscape. As a result, the questions that one can ask are limited by the scale of analysis. An additional concern is that the single geographical scale of GIS technologies impedes our understanding of social and ecological processes that take place at multiple and simultaneous scales.

The Indian and South African case studies demonstrate what Sara McLafferty (2002, p. 268) argues in a different setting—that is, GIS is more than a took kit for spatial analysis: "It encompasses the people who create, utilize and interact with the technology and whose lives are affected by it." The value of Robbins and of McCusker and Weiner's work on participatory GIS is that it points to areas where fruitful connections can be made between political ecology, with its emphasis on power relations and multiscale social and biophysical processes, and GIS mapping. Their identification of the stages of GIS construction suggest areas where resource users can contribute to the production of knowledge that will be to their benefit rather than to their detriment.

Both chapters illustrate the epistemological dimensions of geospatial technologies—how they produce certain types of knowledge and not others. GIS mapping, like all map making, reflects a process of selection and omission. The new mapping technology, like the old, is guided by the objectives of the map maker. Whose interests are served and what political-ecological processes are captured in the GIS is partly based on these choices. "GIS and Society" scholarship encourages us to ask how geospatial technologies might be used to challenge, rather than to reinforce, unequal power relations between experts and resource users, between Western scientific knowledge and local knowledge. They point to future political–ecological research directions that investigate how geospatial technologies (1) structure knowledge and ways of knowing, (2) enhance access to and control over resources, and (3) constrain our understanding of social and biophysical processes that take place at scales other than those inscribed in the image- and databases used to construct GIS layers. Such epistemological and methodological challenges must be negotiated by political ecologists as they continue to ask how the single scaled but multilayered nature of GIS can enrich political ecology and how the multiscaled social and historical approach of political ecology can advance GIS.

North–South Environmental Histories

Future research directions abound in the crossover area where political ecology meets North–South environmental histories as it is covered in the chapters by Sluyter (Chapter 11), Neumann (Chapter 12), and Carney (Chapter 13). These directions are being undertaken in a creative and potentially productive encounter with environmental history and especially its rapidly expanding international focus (Grossman, 1997; Grove, 1995). In the discussion below, each of these directions can be seen as emanating from one of the touchstones of environmental history that overlay ideas of nature, socioeconomic relations with respect to resources and environment, and the biogeophysical realm.

A focus on the translocal and international dimension of cultural ideas of nature is perhaps the most notable of the new directions that is pointed to in the chapters of Sluyter, Neumann, and Carney. Using a framework of geographic interactions in the related environmental histories of the global North and global South, these works shed much new light on such important cultural ideas as the pristine myth, the Edenic myth, and the plantation landscape. Their chapters suggest that the creation and utilization of many of our most influential ideas of nature do not tend to occur strictly within the confines or boundaries of nation-states but rather involve the spatial networks of international relations. Moreover, their North–South environmental histories suggest the importance of an international dimension that includes multisite and translocal relations, in addition to the more familiar metropolitan linkages. Environmental ideas created in these North–South interactions are a powerful and persistent legacy of current conservation and resource management and thus one that offers continued promise in political ecology.

A second future direction is evident in the realm of environment-related social relations. The chapters by Sluyter, Neumann, and Carney clearly establish an emphasis on social relations, which are broadly defined to include political and economic aspects of power in North–South environmental histories. This emphasis is used to demonstrate that differences of race, ethnicity, and economic level are major influences on prevailing cultural ideas of nature and conservation. By undertaking this political–ecological emphasis, Sluyter, Neumann, and Carney provide insights into often overlooked environments—agricultural ones, in large part—and their environmental actors such as African American slaves and indigenous small farmers. Here these chapters signal a promising future direction, for many of the growing number of international conservation histories await a fuller awareness of the environmental role of agriculture and small-scale land users (e.g., those on Mexico and Costa Rica; e.g., Simonian, 1995).

A third future direction that is present in this book's North–South environmental histories is the use of diverse source materials in integrating environmental historical analysis and the evaluation of landscape dynamics. Interest in this direction touches both on broad-ranging philosophical debates and practical and methodological issues. Especially close to geographical political ecology is the budding historical assessment of the role of environmental science in international affairs that are traced through colonialism (Barton, 2001; Grove, 1995). Similarly relevant is the growing discussion of different types of environmental evidence and the rules of proof of scientific reasoning and narrative presentation. This interest is highlighted in such fields as environmental history and restoration, for example, where the integrated historical analysis of human artifacts such as textual documents and settlement remains and ecological scientific methods has gained considerable theoretical and practical import (Egan & Howell, 2001; Foster, 1999).

Innovations are called for in bringing the analysis of biogeophysical conditions and landscape variation more deeply into North–South environmental histories. Clearly Sluyter's chapter suggests that selective integration can combine the analysis of cultural ideas ("cultural transformation") and biogeophysical scientific analysis ("material transformation"). In doing so, Sluyter's chapter identifies a multimethod approach that incorporates the analysis of pollen microfossils, sediment cores, and the critical appraisal of archival documents that refer to landscape and ideas of nature. This methodological approach draws from well-established lineages in cultural ecology and geoarchaeology that exists within and among the geography, anthropology, and paleoenvironmental fields (Butzer & Butzer, 1997; Doolittle, 1995). Promising gains in this direction are also evidenced in the works of Carney and Neumann, for they rely on a sophisticated understanding of the environmental variation of landscapes as central to interpreting the course of North–South histories. For example, the multifaceted argument of Carney's chapter relies on insights from the environmental analysis of the water management landscapes of early rice agriculture.

ADVANCES IN FUTURE DIRECTIONS: CORE THEMES

Nature–Society Fusions

Since the earliest research- and policy-directed formulations of political ecology in the 1980s, the field has placed an interest in nature–society fusions as central to its identity. This focus has continued to expand and has become more theoretically and methodologically sophisticated (Blaikie, 1995, 1998; Forsyth, 2001). At the same time, this evolving interest is paralleled by a proliferating literature in critical geographic thinking about the social–environmental production of nature and resulting socionature (Braun &

Castree, 1998; Castree & Braun, 2001). The discussion below outlines several promising research directions on the theme of fused nature–society analysis that emerge from this book.

Nature–society analysis promises to sharpen the examination of ongoing transitions that are associated with so-called ecological modernization and the growing wave of technoenvironmental and legal regulation (Buttel, 2000). As mentioned earlier, this topic and the paradigm it represents are based on recognition of certain ecoindustrial successes and fairly well-known reforms, such as pollution control, measures for global climate change mitigation, and environmental conservation focused on protected areas. To be sure, ecological modernization is most pronounced in the global North and especially in Northern Europe. Nonetheless, it is marked by strong diffusing tendencies and is being spread through major global and multinational institutions such as the European Union (EU), the United Nations, and the World Bank. The political–ecological processes and outcomes of these institutional influences are most likely to unfold in more geographically uneven and heterogeneous ways than might be thought (Bryant & Bailey, 1997; Watts, 2000). In our view, that unevenness is primed for social–environmental analysis in political ecology.

One example of a combined social–environmental dimension in ecological modernization is the rapidly expanding area of conservation and its emphasis on biodiversity management. In many places, modernization-style effects through ecotourism (Young chapter) and the designation of conservation reserves such as international biospheres (Sundberg and Neumann chapters) have yet to produce the hoped-for effects of either environmental improvement or social benefits for local residents. Conservation-led conditions in many places run counter to the initial phase of across-the-board enthusiasm. New sorts of economic gain, political representation, and identity making are being welded to socioenvironmental change dynamics that do not necessarily portend widespread improvement. At the same time, the findings of these chapters are of growing relevance since they offer contributions that are intended as constructive and ethically necessary for new conservation.

The need for a political–ecological component is also urged in the flurry of laws, policies, and management plans for conservation. Sound interpretation of coupled social–environmental dynamics must inform the debates and discussions that surround the global "conservation boom." Freshly minted scientific regulations of forestry, agriculture, and range ecology are often at odds with the existing political–ecological dynamics of sound land user practices and environmental processes (chapters by Bassett and Koli, Zimmerer, and Turner). These conflicts, which echo debates over the environmental significance of rainforest farming (Naughton-Treves, 2002), gain new weight in the context of expanding conservation policies and regulation. Geographically speaking, an areal dimension seems evermore explicit since environmental conservation and deterioration frequently coexist in close proximity (Daniels

& Bassett, 2002; Zimmerer, 2000). Ecological modernization has already brought particular prominence to geospatial blueprints, such as basins, watersheds, and zones that are being implemented widely in environmental planning and conservation management. These proliferating designs and the foundational role of geospatial technologies such as GIS and remote sensing are well suited to the approach of geographical political ecology.

In general, ecological modernization and conservation are resulting in the enlarged importance of ecological science and derived policy and management concepts. Indeed, expanded environmental and conservation management is heavily dependent on the multifaced contributions of science (Forsyth, 1996). This enlarged role of science, which furnishes both knowledge and claims to political and economic power, is foreshadowed in the modernization of water resource management that took place in the 20th-century history of Spain (Swyngedouw chapter). The emphasis on ecological and environmental sciences seems full of promising inquiries for the theme of nature–society interaction in political ecology. Ethnographic inquiries into environment-related legal and institutional frameworks may be productively combined with the evaluation of geospatial knowledges (St. Martin, 2001). Ecological modernization seems destined to subject every sort of social–environment interaction to an interpretive policy lens and into a prescriptive formulation, which leads to a clarion call for nature–society analysis within political ecology.

Geospatial analysis is no doubt vital to the future of the nature–society theme in geographical political ecology. GIS and remote sensing and such activities as environmental assessment and monitoring can be seen everywhere in the trend toward increased management, regulation, and ecological modernization. In the emphasis on social–environmental interaction that is argued for in this book, the focus on geospatial analysis is likely to offer a scientifically engaged yet critical perspective, for these environment-related technologies are neither inherently bad nor good. It is unfounded to think of political ecology as either purely for or against them. Here the interest in nature–society analysis promises to lead to the judicious incorporation and self-critical awareness of geospatial technologies and analysis, as in the chapter by Robbins and that by McCusker and Wiener. These chapters signal a main thrust in the encounter of geospatial analysis and political ecology, namely, the relations of diverse forms of environmental knowledge that are brought into interplay through ecological modernization efforts such as sustainable development, agrarian reform, and forestry management. This focus builds on the longstanding interest of political ecology and its closely allied fields in local or citizen knowledge. Such knowledge interfaces with the practice and the claims of environmental scientific and technical knowledge and such analogous fields as agricultural science and conservation management (Agrawal, 1995; Bebbington, 1993; Cleveland & Soleri, 2002; Haenn, 1999; Klooster, 2002). Geography as a discipline offers a legacy of productive interest in the relation of these knowledge systems as part of its ongoing discussion of the links among the biogeophysical and human subfields (Schoenberger, 2001; WinklerPrins, 1999). The area of emphasis is also related to the trends in "participatory GIS" that are of immediate relevance to political ecology, which are discussed above.

More specifically, the chapters by Robbins and McCusker and Wiener refer to the geospatial knowledges of forestry and agrarian reform agencies in coexistence that frequently conflict with "nonexpert" spatial–environmental knowledge, thus highlighting the role of "alternative political ecologies." Local or citizen knowledge about geospatial conditions is often fundamental to interpreting political–ecological change. For example, the chapters here demonstrate the strategic relevance of this knowledge to the political claims of disempowered groups vying for customary or just access to resources. The often hidden political–ecological role of local environmental knowledge is exemplified in Carney's work on the almost entirely overlooked agroenvironmental knowledge and contributions that took place in the historic struggles of African American slaves (see also Carney, 1998, 2001). From the viewpoint of political ecology, expanding interest is evident in the geospatial or landscape-based framing of local environmental knowledge. It suggests the idea of ethnolandscape ecologies that form at the interface of local or citizen knowledge studies and technical or scientific analysis (Zimmerer, 2001).

Political Ecology of Scale

From its inception, political ecology has distinguished itself from mainstream nature–society studies (e.g., cultural ecology) by its multiscalar approach to the analysis of human–environmental relationships. Its proponents have long argued for the necessity of going beyond single geographical scale factors influencing land and resource use (e.g., the village) to consider the many regional, national, and international dimensions. For example, one can not explain famine in colonial and postcolonial Nigeria without conceptualizaing the broad scale dynamics of state intervention into rural economies, class relations, the terms of trade, and the accumulation of capital as well as local patterns of gift giving, debt, agro-ecology, and crop yields (Watts, 1983). The focus on simultaneous and multiple processes shaping resource use became a hallmark of political ecology that is exemplified in Blaikie's "chain of explanation" model in his work on the political economy of soil erosion (Balikie, 1985). One of the challenges facing political–ecological scholarship is to break out of these pregiven, scalar containers (local, regional, national, global) to examine human–environmental dynamics that occur at other socially produced and ecological scales. These challenges include being more attentive to the spatiality of social life, especially the politics of scale, and integrating ecological scales into analytical frameworks.

As McCusker and Weiner demonstrate in their chapter, space is not a simple container through which social processes flow. Rather, it is the product of social relationships that assume different configurations under changing conditions. Thus, the challenge for political ecologists is to theorize the social processes producing distinctive geographies of resource access, use, and management. In addition, the spaces of resource control and management are scaled spaces that take a variety of forms. The vertical integration of outgrowers in a contract farming scheme (Little & Watts, 1994) produces a hierarchical form of resource control extending from the transnational contracting firm to the local field scale. Water flows and management regimes linking rural and urban areas create horizontal scalar processes, as demonstrated by Swyngedouw and Pelling in their chapters. As Swyngedouw (1997, p. 141) argues, "The theoretical and political priority . . . never resides in a particular geographical scale, but rather in the process through which particular scales become (re)constituted."

The politics of scale literature emphasizes the *social* production of scale. The challenge of political ecology is to integrate the scales of biophysical dynamics into our research frameworks and policy discussions. Future political ecological research might consider how ecological scale interacts with socially constructed scales to produce distinctive environmental geographies. Four productive avenues of research are suggested here: (1) the scales of ecological dynamics; (2) functional conservations areas; (3) mismatches between ecological and social scales; and (4) fragmented scales.

The scales of ecological processes are poorly understood by ecologists and conservation biologists. Too often, the spatial dynamics of species and ecosystems are arbitrarily defined to fit within human-designed management areas or researcher study plots rather than by their functional requirements (Wiens, 1989; Hobbs, 1998). Yet the scale of observation can influence how landscape patterns and ecological processes are detected (Wiens, 1989, p. 386). For example, the relative importance of equilibrium or non-equilibrium processes in a given landscape will vary according to the areal extent of a study. Thus, determining the spatial extent of a species or ecosystem and the appropriate size of the unit of observation are critical to understanding ecological processes. To convey the importance of spatial scale to ecology, ecologists have coined the term "ecological neighborhood" to refer to spaces that are scaled to specific ecological processes (Wiens, 1989, p. 390).

Poiani et al. (2000) integrate spatial scaling into their theoretical framework for the design of nature reserves and conservation areas. They propose conserving ecosystems and species within functional conservation sites, landscapes, and networks at four geographical scales: local, intermediate, coarse, and regional. For example, local-scale functional conservation areas would protect small patch ecosystems (outcrops, unique soils) and local-scale species in a fairly restricted area, whereas intermediate-scale conservation areas would comprise large patch ecosystems (e.g., salt marshes) and intermediate-scale species. The point is that the functional area of the species or ecosystem should determine the scale of the conservation area. Too often, management units are based on arbitrary administrative organizational levels (national, regional, district) or protected area designs, such as the concentric ring pattern of biosphere reserves, rather than on the ecological processes and scales that sustain ecosystems and species.

Scale mismatches occur where the spatial requirements of a species or ecosystem do not correspond with administrative levels of management. The case studies of Sundberg and Young in this volume demonstrate this point. Along the Baja coast of Mexico, the state designates zones where cooperatives have exclusive rights to fish or guide tourist boats to watch whales. However, the community-scale resource management organization does not match the transient resource (fish, whales) whose (oceanic) scale is much broader. This scale mismatch hinders the effective management of the resources whose broad scale requirements demand international as well as local resource management regimes. In the case of the Petén peninsula of Guatemala, NGOs and the state seek to control resource use in a demarcated area that contains neither a sedentary population nor a closed economy. Population mobility and economic activities transcend the territorial limits of the nature preserve. In this case, the scale of human activities do not match that of the biosphere reserve.

Political ecologists might also be attentive to fragmented scales such as landscapes divided by roads, fields, or political boundaries which affect the spatial extent of ecological processes. These socially produced scales disrupt the movement of species such as migratory birds whose spatial ranges vary by species and are susceptible to environmental changes such as global warming (Parmesan & Yohe, 2003). Fragmented landscapes can also alter the effects (spatial extent, intensity) of certain ecological process such as fire on the ratio of grasses and trees in savanna landscapes (Wiens, 1995; see also Bassett & Koli, this volume).

The hallmark of the political ecology of scale is the simultaneous consideration of social and biophysical processes that produce distinctive sociospatial configurations of resource use. Matthew Turner's analysis of grazing-induced rangeland degradation in Sahelian West Africa joins the spatial-temporal distribution of rainfall to changing social relations of production and the effects of European policies on beef exports. This multiscalar approach leads him to assess the limits of conventional environmental science (e.g., carrying capacity, stocking rates) in our understanding of the political ecology of overgrazing. In his discussion of the spaces of biodiversity in the Peruvian Andes, Zimmerer (this volume) examines the simultaneous social and biophysical processes shaping crop patterns. His meso-scale analysis reveals a "crazy quilt" of overlapping farm spaces that contradicts the hierarchical model of vertical zonation. Both studies integrate the ecological and social dimensions of land use in ways that advance our understanding of the production of space (farm spaces, spatial patterns of grazing pressure) and their associated scales (overlapping patchworks, regional networks) that contrast with the more conventional and misconceived spaces and scales of land use (vertical zonation, stocking rates). The challenge and strength of political ecology is this creative delimitation of the spaces and scales of resource management that are both the medium and the outcome of intertwined social-environmental dynamics.

NOTE

1. The Rio Earth Summit 1992; the Johannesburg World Summit on Sustainable Development 2002.

REFERENCES

Agrawal, A. (1995). Dismantling the divide between indigenous and scientific knowledge. *Development and Change, 26*, 413–439.

Barton, G. (2001). Empire forestry and the origins of environmentalism. *Journal of Historical Geography, 27*(4), 529–552.

Bassett, T. J. (2002). Women's cotton and the spaces of gender politics in northern Côte d'Ivoire. *Gender, Place, and Culture, 9*(4), 351–370.

Bassett, T. J., & Crummey, D. (2003). *African savannas: Global narratives and local knowledge of environmental change.* Oxford, UK, and Portsmouth, NH: James Currey & Heinemann.

Bebbington, A. (1993). Modernization from below: An alternative indigenous development? *Economic Geography, 69*(3), 274–92.

Beck, U. (1992). *Risk society: Towards a new modernity* (M. Ritter, Trans.). Newbury Park, CA: Sage.

Beck, U. (1999). *World risk society.* Cambridge, UK: Polity Press.

Blaikie, P. (1985). *The political economy of soil erosion in developing countries.* Essex, UK: Longman.

Blaikie, P. (1995). Changing environments or changing views?: A political ecology for developing countries. *Geography, 80*(3), 203–214.

Blaikie, P. (1998). A review of political ecology: Issues, epistemology, and analytical narratives. *Zeitschrift für Wirtschaftsgeographie, 3–4*, 131–147.

Braun, B., & Castree, N. (Eds.). (1998). *Remaking reality: Nature at the millennium.* London: Routledge.

Bryant, R., & Bailey, S. (1997). *Third world political ecology.* London: Routledge.

Buttel, F. H. (2000). Ecological modernization as social theory. *Geoforum, 31*, 57–65.

Butzer, K. W., & Butzer, E. K. (1997). The "natural" vegetation of the Mexican Bajío: Archival documentation for a 16th-century savanna environment. *Quaternary International, 43–44,* 161–172.

Carney, J. A. (1998). The role of African rice and slaves in the history of rice cultivation in the Americas, *Human Ecology, 26,* 525–545.

Carney, J. A. (2001). *Black rice: The African origins of rice cultivation in the Americas.* Cambridge, MA: Harvard University Press.

Castle, E. N. (2002). Social capital: An interdisciplinary concept. *Rural Sociology, 67*(5), 331–349.

Castree, N., & Braun, B. (Eds.). (2001). *Social nature: Theory, practice, and politics.* Malden, MA: Blackwell.

Coomes, O. (1996). State credit programs and the peasantry under populist regimes: Lessons from the APRA experience in Peruvian Amazonia. *World Development, 24*(8), 1333–1346.

Coomes, O., & Barham, B. (1997). Rain forest extraction and conservation in Amazonia. *Geographical Journal, 163*(2), 180–188.

Cormier-Salem, M-C., Juhé-Beaulaton, D., Boutrais, J., & Roussel, B. (Eds.). (2002). *Patrimonialiser la nature tropicale: Dynamiques, locales, enjeux inernationaux.* Paris: IRD.

Daniels, R., & Bassett, T. J. (2002). The spaces of conservation and development around Lake Nakuru National Park, Kenya. *Professional Geographer, 54*(4), 481–490.

DiChiro, G. (1998). Nature as community: The convergence of environment and social justice. In M. Goldman (Ed.), *Privatizing nature: Political struggles for the global commons* (pp. 120–143). New Brunswick, NJ: Rutgers University Press.

Doolittle, W. (1995). Indigenous development of Mesoamerican irrigation. *Geographical Review, 85,* 301–323.

Egan, D., & Howell, E. A. (Eds.). (2001). *The historical ecology handbook: A restorationists' guide to reference ecosystems.* Washington, DC: Island Press.

Forsyth, T. (1996). Science, myth and knowledge: Testing Himalayan environmental degradation in Thailand. *Geoforum, 27*(3), 375–392.

Forsyth, T. (2000). Critical realism and political ecology. In J. Lopez and G. Potter (Eds.), *After postmodernism: An introduction to critical realism* (pp. 146–154). London: Athlone.

Foster, D. R. (1999). *Thoreau's country: Journey through a transformed landscape.* Cambridge, MA: Harvard University Press.

Friedman, S. S. (2001). Academic feminism and interdisciplinarity. *Feminist Studies, 27*(2), 504–510.

Fry, G. L. A. (2001). Multifunctional landscapes: Towards transdisciplinary research. *Landscape and Urban Planning, 57*(3–4), 159–168.

Goldman, M. (2002). Notes from the World Summit in Johannesburg: History in the making? *Capitalism, Nature, Socialism 13*(4), 68–79.

Golledge, R. G. (2002). The nature of geographic knowledge. *Annals of the Association of American Geographers, 92*(1), 1–14.

Grossman, L. S. (1997). Soil conservation, political ecology, and technological change on Saint Vincent. *Geographical Review, 87*(3), 353–374.

Grove, R. H. (1995). *Green imperialism: Colonial expansion, tropical island edens, and the origins of environmentalism, 1600–1860*. Cambridge, UK: Cambridge University Press.

Hobbs, R. J. (1998) Managing ecological systems and processes. In D. Peterson & V. T. Parker (Eds). *Ecological scale: Theory and applications* (pp. 459–484). New York: Columbia University Press.

Homewood, K., Lambin, E. F., Coast, E., Kariuki, A., Kikula, I., Kivela, J., et al. (2001). Long-term changes in Serengeti–Mara wildebeest and land cover: Pastoralism, population, or policies? *Proceedings of the National Academy of Science, 98*(22), 12544–12549.

Jackson, C. (2002). Disciplining gender? *World Development, 30*(3), 497–509.

Jarosz, L., & Qazi, J. (2000). The geography of Washington's world apple: Global expressions in a local landscape. *Journal of Rural Studies, 16*(1), 1–11.

Hurley, A. (1995). *Environmental inequalities: Class, race, and industrial pollution in Gary, Indiana, 1945–1980*. Chapel Hill: University of North Carolina Press.

Klooster, D. (2002). Toward adaptive community forest management: Integrating local forest knowledge with scientific forestry. *Economic Geography, 78*(1), 43–70.

Klooster, D., & Masera, O. (2000). Community forest management in Mexico: Carbon mitigation and biodiversity conservation through rural development. *Global Environmental Change, 10*, 259–272.

Lambin, E. F., Turner, B. L., Geist, H., Agbola, S. B., Angelsen, A., Bruce, J. B., et al. (2001). The causes of land-use and land-cover change: Moving beyond the myths. *Global Environmental Change, 11*, 261–269.

Laurie, N., Andolina, R., & Radcliffe, S. (2002). The excluded "indigenous"?: The implications of multi-ethnic policies for water reform in Bolivia. In R. Sider (Ed.), *Multiculturalism in Latin America: Indigenous rights, diversity, and democracy* (pp. 252–276). New York: Palgrave.

Martinez-Alier, J. (2000). Retrospective environmentalism and environmental justice movements today. *Capitalism, Nature, Socialism, 11*(4), 45–50.

McCarthy, J. (2002). First World political ecology: Lessons from the Wise Use Movement. *Environment and Planning A, 34*(7), 1281–1302.

McLafferty, S. (2002). Mapping women's worlds: Knowledge, power and the bounds of GIS. *Gender, Place and Culture, 9*(3), 263–269.

Moore, D. (1993). Contesting terrain in Zimbabwe's eastern highlands: Political ecology, ethnography, and peasant resource struggles. *Economic Geography, 69*(4), 380–401.

Mutersbaugh, T. (2002a). Migration, common property, and communal labor: Cultural politics and agency in a Mexican village. *Political Geography, 21*, 473–494.

Mutersbaugh, T. (2002b). "The number is the beast": A political economy of transnational certified organic coffee and coffee producer unionism. *Environment and Planning A, 34*, 1165–1184.

Naughton-Treves, L. (2002). Wild animals in the garden: Conserving wildlife in Amazonian agroecosystems. *Annals of the Association of American Geographers, 93*(2), 488–506.

Neumann, R. (1998). *Imposing wilderness: Struggles over livelihood and nature preservation in Africa*. Berkeley and Los Angeles: University of California Press.

Neumann, R. (2001). Disciplining peasants in Tanzania: From state violence to self-surveillance in wildlife conservation. In N. Peluso & M. Watts (Eds.), *Violent environments* (pp. 305–327). Ithaca, NY: Cornell University Press.

Parmesan, C., & Yohe, G. (2003). A globally coherent fingerprint of climate change impacts across natural systems. *Nature, 42*, 32–42.

Peluso, N. (1996). Coercing conservation?: The politics of state resource control. *Global Environmental Change, 3*(2), 199–218.

Pezzoli, K. (1997). Sustainable development: A transdisciplinary overview of the literature. *Journal of Environmental Planning and Management, 40*(5), 549–574.

Poiani, K., Richeter, B. D., Anderson, M. G., & Richter, H. E. (2000). Biodiversity conservation at multiple scales: Functional sites, landscapes, and networks. *BioScience, 50*(2), 133–146.

Rayner, S. (2003, March 12). *Interview.* Gaylord Nelson Institute for Environmental Studies, University of Wisconsin, Madison.

Rocheleau, D., Thomas-Slayter, B., & Wangari, E. (Eds.). (1996). *Feminist political ecology: Global issues and local experience.* London: Routledge.

St. Martin, K. (2001). Making space for community resource management in fisheries. *Annals of the Association of American Geographers, 91*(1), 122–142.

Schoenberger, E. (2001). Interdisciplinarity and social power. *Progress in Human Geography, 25*(3), 365–382.

Schroeder, R. (2000). Beyond distributive justice: Environmental justice and resource extraction. In C. Zerner (Ed.), *People, plants, and justice: Resource extraction and conservation in tropical developing countries* (pp. 52–64). New York: Columbia University Press.

Simonian, L. (1995). *Defending the land of the jaguar: A history of conservation in Mexico.* Austin: University of Texas Press.

Stonich S. C. (1995). The environmental quality and social justice implications of shrimp mariculture development in Honduras. *Human Ecology, 23*(2), 143–168.

Swyngedouw, E. (1997). Neither global nor local: "Glocalization" and the politics of scale. In K. R. Cox (Ed.), *Spaces of globalization: Reasserting the power of the local* (pp. 137–166). New York: Guilford Press.

Turner, B. L., II. (2002). Contested identities: Human–environment geography and disciplinary implications in a restructuring academy. *Annals of the Association of American Geographers, 92*(1), 52–74.

Turner, B. L., Villar, S. C., Foster, D., Geoghegan, J., Keys, E., Klepeis, P., et al. (2001). Deforestation in the southern Yucatán peninsular region: An integrative approach. *Forest Ecology and Management, 154*, 353–370.

Turner, M. (1999). Merging local and regional analyses of land-use change: The case of livestock in the Sahel. *Annals of the Association of American Geographers, 89*(2), 191–219.

Vayda, A. P., & Walters, B. B. (1999). Against political ecology. *Human Ecology, 27*(1), 167–179.

Watts, M. (1994). Development II: The privatization of everything. *Progress in Human Geography, 18*, 371–384.

Watts, M. J. (1983). *Silent violence: Food, famine and peasantry in northern Nigeria.* Berkeley and Los Angeles: University of California Press.

Watts, M. J. (1998). Nature as artifice and artifact. In B. Braun & N. Castree (Eds.), *Remaking reality: Nature at the millenium* (pp. 243–269). London: Routledge.

Watts, M. J., & Goodman, D. (1997). Agrarian questions: Nature, culture, and industry in fin-de-siècle agro-food systems. In D. Goodman & M. J. Watts (Eds.), *Globalising food: Agrarian questions and global restructuring* (pp. 1–34). London: Routledge.

Watts, M. W. (2000). Political ecology. In E. Sheppard & T. J. Barnes (Eds.), *A companion to economic geography* (pp. 257–274). Malden, MA: Blackwell.

Wiens, J. A. (1989). Spatial scaling in ecology. *Functional Ecology, 3,* 385–397.

Wiens, J. A. (1995). Landscape mosaics and ecological theory. In L. Hansson, L. Fahrig, & G. Merriam (Eds.), *Mosaic landscapes and ecological processes* (pp. 1–26). London: Chapman & Hall.

WinklerPrins, A. M. G. A. (1999). Local soil knowledge: A tool for sustainable land management. *Society and Natural Resources, 12,* 151–161.

Zerner, C. (2002). Introduction: Toward a broader vision of justice and nature conservation. In C. Zerner (Ed.), *People, plants and justice: The politics of nature conservation* (pp. 3–20). New York: Columbia University Press.

Zimmerer, K. S. (1996). Ecology as cornerstone and chimera in human geography. In C. Earle, K. Mathewson, & M. Kenzer (Eds.), *Concepts in human geography* (pp. 161–188). London: Rowman & Littlefield.

Zimmerer, K. S. (2000). Re-scaling irrigation in Latin America: The cultural images and political ecology of water resources. *Ecumene, 7*(2), 150–175.

Zimmerer, K. S. (2001). The common ground of geography and the new ethnobiology: Links to a framework of ethno-landscape ecology. *Geographical Review, 91*(4), 725–734.

Index

About the Editors

Karl S. Zimmerer is Professor in the Department of Geography, the Institute for Environmental Studies, the Program in Development Studies, and the Land Tenure Center at the University of Wisconsin–Madison. He is Co-Director of the Environment-and-Development Advanced Research Circle (EDARC). Dr. Zimmerer received his PhD in geography at the University of California at Berkeley, and has received numerous honors and awards, including the John Simon Guggenheim Fellowship and the Carl O. Sauer and H.C. Cowles Awards of the Association of American Geographers. Dr. Zimmerer's research includes a project on the political ecology of global development organizations, agrobiodiversity conservation, and environmental change among Quechua and non-Quechua farmers in Peru and Bolivia. He is also researching the seed networks and institutional and NGO support frameworks for growers of maize and potatoes in Mexico and the Andean countries. Currently, Dr. Zimmerer is undertaking a study of the idea of ecoregional planning in Latin America, the United States, and Europe and the roles of agriculture, international science, and globalization. His previous publications include *Changing Fortunes: Biodiversity and Peasant Livelihood in the Peruvian Andes* and *Nature's Geography: New Lessons for Conservation in Developing Countries*.

Thomas J. Bassett is Professor of Geography at the University of Illinois at Urbana–Champaign. He received his PhD in geography at the University of California at Berkeley. For more than 20 years, Dr. Bassett has been engaged in long-term research in Côte d'Ivoire on land use and land cover change in the savanna region, with emphasis on the social and agricultural history of cotton, Fulβe livestock raising, land rights systems, and market hunting. He is the author of *The Peasant Cotton Revolution in West Africa: Côte d'Ivoire, 1880–1998* and the coauthor of *African Savannas: Global Narratives and Local Knowledge of Environmental Change, Mapping Africa to 1900,* and *Land in African Agrarian Systems.* Dr. Bassett's current research focuses on the political–ecological dynamics of land privatization in northern Côte d'Ivoire.

Contributors

Thomas J. Bassett, PhD, Department of Geography, University of Illinois, Urbana–Champaign, Illinois

Judith Carney, PhD, Department of Geography, University of California, Los Angeles, California

Koli Bi Zuéli, Doctorat, Institut de Géographie Tropicale, Université de Cocody, Abidjan, Côte d'Ivoire

Brent McCusker, PhD, Department of Geology and Geography, West Virginia University, Morgantown, West Virginia

Roderick P. Neumann, PhD, Department of International Relations, Florida International University, Miami, Florida

Mark Pelling, PhD, Department of Geography, University of Liverpool, Liverpool, England

Paul Robbins, PhD, Department of Geography, Ohio State University, Columbus, Ohio

Andrew Sluyter, PhD, Department of Geography and Anthropology, Louisiana State University, Baton Rouge, Louisiana

Juanita Sundberg, PhD, Department of Geography, University of British Columbia, Vancouver, Canada

Erik Swyngedouw, PhD, School of Geography and the Environment and St. Peter's College, Oxford University, Oxford, England

Matthew Turner, PhD, Department of Geography, University of Wisconsin–Madison, Madison, Wisconsin

Daniel Weiner, PhD, Department of Geology and Geography, West Virginia University, Morgantown, West Virginia

Emily H. Young, PhD, San Diego Foundation Environment Program, San Diego, California

Karl S. Zimmerer, PhD, Department of Geography, University of Wisconsin–Madison, Madison, Wisconsin

5.694